All the Multiverse!

Starships Exploring the Endless Universes of the Cosmos Using the Baryonic Force

Once more unto the breach, dear friends!
All the Universes are within Man's Reach![1]

---

[1] First line: *King Henry V*, Will Shakespeare; second line: the author.

All the Multiverse!

Starships Exploring the Endless Universes of the Cosmos Using the Baryonic Force

STEPHEN BLAHA

BLAHA RESEARCH

ISBN: 978-0-9893826-3-2

Cover Credits
Symbolic view of the Multiverse in "baryonic light" – not electromagnetic light containing an array of black, antimatter-driven starships. Dark spots represent universes. Copyright © Stephen Blaha 2013-2014. All Rights reserved.

Rev. 00/00/01   April 10, 2014

To My Grandchildren:

Milan, Alexandre, Maxim, and Nicholas

Some Other Books by Stephen Blaha

From Asynchronous Logic to The Standard Model to Superflight to the Stars (Blaha Research, Auburn, NH, 2011)

From Asynchronous Logic to The Standard Model to Superflight to the Stars; Volume 2: Superluminal CP and CPT, U(4) Complex General Relativity and The Standard Model, Complex Vierbein General Relativity, Kinetic Theory, Thermodynamics (Blaha Research, Auburn, NH, 2012)

The Algebra of Thought & Reality: The Mathematical Basis for Plato's Theory of Ideas, and Reality Extended to Include A Priori Observers and Space-Time; Second Edition (Pingree-Hill Publishing, Auburn, NH, 2009)

Quantum Big Bang Cosmology: Complex Space-time General Relativity, Quantum Coordinates, Dodecahedral Universe, Inflation, and New Spin 0, ½, 1 & 2 Tachyons & Imagyons™ (Pingree-Hill Publishing, Auburn, NH, 2004)

SuperCivilizations: Civilizations as Superorganisms (McMann-Fisher Publishing, Auburn, NH, 2010)

Standard Model Symmetries, And Four and Sixteen Dimension Complex Relativity; The Origin Of Higgs Mass Terms (Blaha Reasearch, Auburn, NH, 2012)

The Bridge to Dark Matter; A New Sister Universe; Dark Energy; Inflatons; Quantum Big Bang; Superluminal Physics; An Extended Standard Model Based on Geometry (Blaha Reasearch, Auburn, NH, 2013)

Universes and Multiverses: From a New Standard Model to a Physical Multiverse; The Big Bang; Our Sister Universe's Wormhole; Origin of the Cosmological Constant, Spatial Asymmetry of the Universe, and its Web of Galaxies; A Baryonic Field between Universes and Particles; Flatverse Extended Wheeler-DeWitt Equation (Blaha Reasearch, Auburn, NH, 2014)

Available on bn.com, Amazon.com, Amazon.co.uk and other international web sites as well as at better bookstores (through Ingram Distributors).

Preface

Recently, we have shown in a series of books that an Extended Standard Model of Elementary Particles can be derived from Asynchronous Logic combined with *complex* four-dimensional space-time geometry. We have also shown that a *complex* sixteen-dimensional multiverse consisting of universes embedded in a flat space, with each universe satisfying a complex Wheeler-Dewitt equation, leads to an understanding of many features seen in our universe: the origin of the Cosmological Constant, the spatial asymmetry of our universe, and the web of Galaxies recently found in our universe. It also suggests a wormhole connection to a sister universe and the existence of a baryonic field between universes and particles (Blaha (2014)). *The combination of these successes in explaining both the microscopic and cosmic features of our universe, and resolving quantum issues of the Wheeler-DeWitt equation, strongly suggests our theory is in accord with Nature. It gives us a reason to look into the distant future with confidence.*

The multiverse concept has recently received significant support with the discovery of cosmic inflation and the existence of primordial gravitational waves (BICEP2 Collaboration announcement March 17, 2014). These discoveries strongly suggest that a multiverse exists in the cosmos – in part created by space-time universe bubbles generated by the rapid irregular expansion of our universe. Thus the multiverse of Blaha (2014) is strengthened.

In tens of thousands of years Man may reach beyond our universe to countless other universes located in the space beyond our universe that we call the Multiverse, if Man advances intellectually, socially, and technologically. Our multiverse is an infinite 16-dimensional flat space that we call the Flatverse. We see reason to believe that an infinity of universes, including our own universe, may exist within the Flatverse separated generally by large distances of perhaps trillions of light years – island universes containing matter and energy. The all-enveloping Flatverse is like a desert – no matter and no energy of the types with which we are familiar – with universes dotting the Flatverse like oases. The book's cover shows a symbolic view of the Multiverse of universes.

This book makes a leap of tens of thousands of years of research and development – perhaps 50,000 years (four times the approximately 12,500 year period from human hunter-gatherer clans to the present) – to describe travel to far universes from our universe. We will describe the Multiverse as we conceive it. Then we will describe the general features of a starship for universe travel. We will call it a *uniship*. The features of the uniship will differ significantly from a starship such as our previously suggested quark-gluon ion drive starship for travel within our universe. (See appendix A for a description of our starship design and features.) The uniship will require a more powerful, more economical drive based on particle-antiparticle annihilation (Appendix B). This drive must enable travel in fifteen different directions in the Flatverse. (Multi-dimensional drives have not been seriously considered previously.) We propose a form of multi-dimensional drive.

The uniship will require radically different mechanisms for seeing and navigating within the Flatverse. The mechanisms will have to accommodate using our 3-dimensional eyes to see and navigate in the 15-dimensional Flatverse space. We propose mechanisms for these purposes based on a fifth force of nature: a baryonic force that was suggested over sixty years ago. We show this force is embodied in a 15-dimensional field similar to the electromagnetic field. As the electromagnetic field enables us to see and navigate in three dimensions, so the fifteen dimensional baryonic field gives us eyes in fifteen dimensions.

Another issue is the entry and exit of universes. We consider entry and exit in some detail due to its dynamic and topological ramifications. Changing directions and dimensions are significant problems.

The discussions in this book are the culmination of fourteen years of research on faster than light motion, The Standard Model of Elementary Particles, Gravitation, Cosmology and our Multiverse of universes. These studies appear in numerous books by the author.

In writing this book we presume to look to the distant future and make assumptions that are reasonable but not guaranteed. The most significant assumption is the existence of a fifth force – a baryonic force – that makes travel out of our universe possible and plays a major role in travels in the multiverse. This assumption is supported by theoretical evidence – the conservation of baryon number. The second most significant assumption is the existence of the multiverse of universes. The existence of other universes and thus a multiverse is supported by the need for a mass for the Higgs Mechanism, the need for a quantum observer, and the need for a clock for the universe. Our sister universe, and the other universes, meet these needs. If we have evidence for two universes it is but a "small" leap to a multitude of universes – the Multiverse –

and a Flatverse to fill the gap between them. The likelihood of these assumptions, and the novel, new perspectives they lead to, caused this author to proceed to explore the possibilities of emerging from our universe and traveling to other universes knowing that it would not be feasible for many tens of thousands of years.

The presentation in this book is largely self-contained. The reader may wish to also read some recent books by the author of special relevance for the present work:

Universes and Multiverses: From a New Standard Model to a Physical Multiverse; The Big Bang; Our Sister Universe's Wormhole; Origin of the Cosmological Constant, Spatial Asymmetry of the Universe, and its Web of Galaxies; A Baryonic Field between Universes and Particles; Flatverse Extended Wheeler-DeWitt Equation

The Origin Of Higgs Mass Terms; and *The Bridge to Dark Matter; A New Sister Universe; Dark Energy; Inflatons; Quantum Big Bang; Superluminal Physics; An Extended Standard Model Based on Geometry*

From Asynchronous Logic to The Standard Model to Superflight to the Stars; Volume 2: Superluminal CP and CPT, U(4) Complex General Relativity and The Standard Model, Complex Vierbein General Relativity, Kinetic Theory, Thermodynamics;

From Asynchronous Logic to The Standard Model to Superflight to the Stars;

Standard Model Symmetries, and Four and Sixteen Dimension Complex Relativity;

After Man has explored the stars, has explored the galaxies of our universe, there will still be the quest to explore the many universes of the Cosmos: to see eternity's sunrise, to reach the heights and depths of fundamental Reality, and so to grow to maturity as a species. And yet there will be time – time to further explore mind and matter. In this book we will only talk of space, time and matter. A deeper subject – universes of the mind – remains to be explored to seek what immortal hand or mind could frame these "fearful" symmetries of Nature.

CONTENTS

FIGURES and TABLES

1   The Multiverse of Universes

We are much concerned about our universe – its history, its future, its features, and bizarre topological structures that might exist. These studies, originating in astronomical observation and General Relativity, have given us black holes[2] (now found), wormholes (not found as yet) and a preliminary understanding of the evolution of stars, galaxies, super galaxies and other features.

Recently, we have shown in a series of books that an Extended Standard Model of Elementary Particles can be derived from Asynchronous Logic combined with *complex* four-dimensional space-time geometry. We have also shown that a *complex* sixteen-dimensional multiverse consisting of universes embedded in a flat space, with our universe satisfying a complex Wheeler-Dewitt equation, leads to an understanding of many features of our universe: the origin of the Cosmological Constant, the spatial asymmetry of our universe, and the web of Galaxies recently found in our universe as well as suggesting a wormhole connection to a sister universe and the existence of a baryonic field between universes and particles (Blaha (2014)). *The combination of these successes in explaining both the microscopic and cosmic features of our universe, and resolving quantum issues of the Wheeler-DeWitt equation, strongly suggests our theory is in accord with Nature. The new BICEP2 Collaboration data (March 17, 2014) provides tentative support for the existence of a multiverse – a set of universes existing in a flat space.[3] It gives us a reason to look into the distant future with confidence.*

This book builds on our new view of this universe, and other universes, as described in Blaha (2014). This chapter summarizes a large part of that book as a preliminary to the description of a strange new type of space craft with many novel features for travel outside our universe. We do not expect this craft to be

---

[2] Professor Hawking has recently put forward the notion that black holes are in fact "grey" in the sense that they can emit as well as absorb.

[3] The standard view of a multiverse, held by many, is that it originated in the Big Bang and subsequent inflationary expansion of our universe. It appears to be extremely anthropocentric to think our universe began it all – the multiverse. Rather it is more reasonable to think our universe – however it arose – is but one of many universes. Our universe is not the center of the multiverse but merely one of its constituents.

built for many millennia. But if Man continues to progress, and there is some strong doubt that Man will, we might then expect trans-universe craft to be built perhaps within the next fifty thousand years.

We shall call these craft *uniships*[4] since they will travel between universes.

One might ask why we would consider uniships since they are in the distant future at best. After some thought this author thought it might help set Man's sights higher and, to a small degree, guide research efforts towards that high goal. This universe will eventually prove too small for Man's progress. We shall pursue this line of thought further in chapter 2.

1.1 What is Our Multiverse?

The multiverse, as we conceive it, is a complex, 16-dimensional, flat Euclidean space within which there are island universes.[5] There is no matter or energy in the flat space between universes. The totality of the 16-dimensional space including the island universes within it we call the *Flatverse*. The entire Flatverse has a quantum vacuum state containing fermions and bosons as described in conventional quantum field theory.

A universe occupies a region within the Flatverse. Its gravitational field and mass-energy are confined to the universe's region – rather like quarks and gluons are confined within hadrons.[6] We will consider the confinement of gravitation in more detail later. We will also call universes, *island universes*, since they are localized regions with a curved metric occurring within the Flatverse.

In Flatverse coordinates a universe can change its location and/or size. Universes can expand or contract. Within a universe, its size and mass-energy distributions may change. We will consider a topologically simple multiverse with at most wormholes connecting universes or parts of universes.

Island universes are universes containing matter and energy and thus have gravitational fields that give them a curved space-time structure. The total

---

[4] Uniship is pronounced "you-nee-ship." It is an abbreviation for universe-traversing starship.

[5] Prior definitions of the multiverse usually assume one universe – our universe – within which topological structures may appear. DeWitt and others have developed theories of quantum gravity within this universe. We will consider aspects of their work later.

[6] The possibility that the gravitational field and the strong interaction gauge field might have similarities such as confinement and a linear potential was first considered by this author in "Quantum Gravity and Quark Confinement" by Stephen Blaha, Lett. Nuovo Cim. 18:60, 1977. Honorable Mention in the Gravity Research Foundation Essay Competition in 1978.

energy of an island universe is zero – the energy of the particles is canceled by the energy of the gravitational field. Thus the gravity field is zero outside of the island universes although the Flatverse vacuum has a quantum graviton part. The uniformity of the graviton vacuum precludes classical gravitational interactions between island universes.[7]

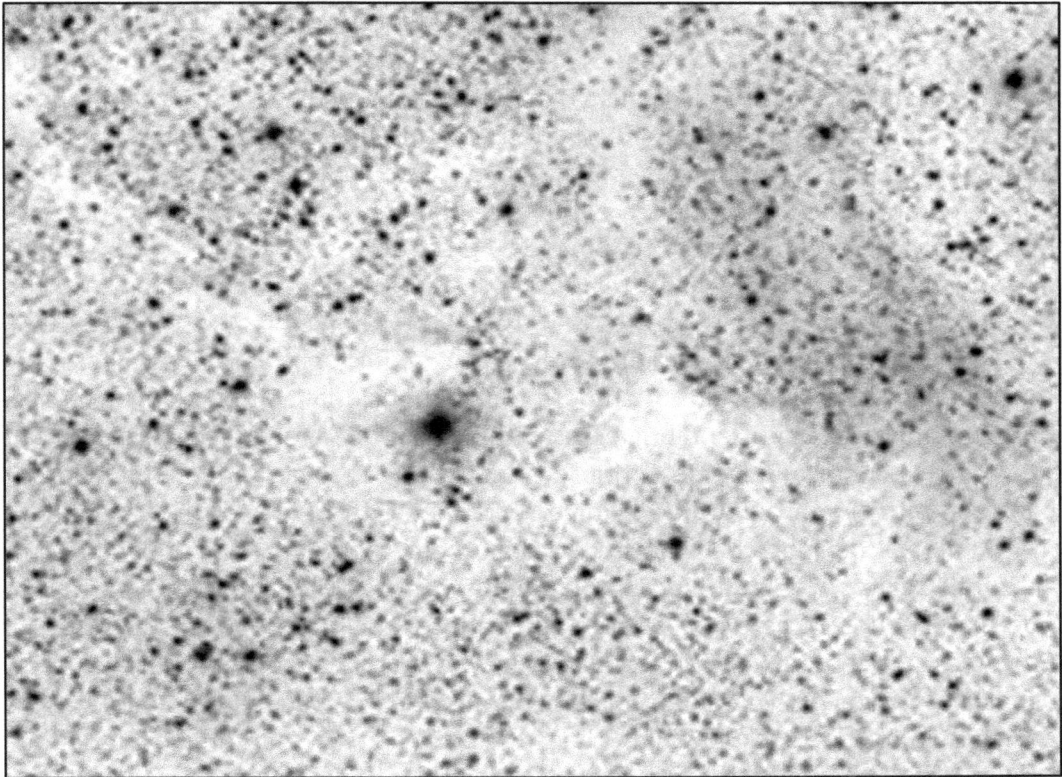

Figure 1.1. A multiverse visualization of the Flatverse representing universes as black spots.

Island universes have a non-zero baryon number judging from the baryon number of our universe. All other quantum numbers total to zero. We have

---

[7] The Flatverse outside of universes does not have matter or energy or it would not be flat. However it does support the known forces of nature in the Standard Model plus Gravitation. These interactions do not have sources in the Flatverse. So only their vacuum is dynamic.

suggested in Blaha (2014) that island universes may have a form of "spin" that does not necessarily relate to the interior physical properties of the universe just as the spin of a fundamental particle such as an electron does not relate to the motion of the electron's "internals."

Because baryon number is conserved exactly as far as we know we propose baryon number conservation is associated with an ultra-weak force embodied in a spin 1 gauge field with a massless particle that we call the *planckton.* The form of its dynamics is similar to quantum electrodynamics. This force acts between baryon particles and between universes due to their total baryon "charge." The vacuum fluctuations of this field can generate universe-antiuniverse pairs. Antiuniverses have "negative" baryon number. We view universes/antiuniverses, when created (Big Bangs), as ultra-small particles of mass-energy that can proceed to expand to great size. If they are generated by a vacuum fluctuation they will, after possibly a certain time, recombine into the vacuum.

While topologically complex universes are quite possible, we will assume that island universes are closed and are four-dimensional complex universes like our own universe. Flat island universes within the multiverse are possible.

The size of a universe within the multiverse can vary and can expand or contract with time. The scale of universes appears to range from very small to many billions of light years based on our understanding of our universe. The "natural" scale of the multiverse would appear to be trillions of light years since one expects universes to have large separations much greater than the size of universes in general. Universes can also be fairly close together, as well, especially since universe-antiuniverse creation via vacuum fluctuations is possible in our theory through the fifth force – the baryonic gauge field. Universes and antiuniverses so generated would initially be very small and very close together at their "Big Bang" point.

1.2 Evidence for the Multiverse

At first glance it would seem impossible to produce evidence for the existence of other universes. For if we could "see" electromagnetically or gravitationally another universe why should we not consider it a part of our universe?

However, there are a number of indirect ways of showing that current theory and experiment require the existence of at least one other universe. And if such an additional universe exists, then additional dimensions would be required (four?), and intervening space between the universes must exist. Such a space must be flat since the universes are gravitationally self-contained. If there were random matter/energy between the universes, then they would constitute a third universe.

The first indication of the need for something more is the presence of mass terms in the Higgs' particle dynamic equations. The coupling constants of the quartic Higg's terms are dimensionless. The Higgs mechanism reduced the source for all the mass terms of the other elementary particles to an origin in the Higgs dynamics sector. But a further reduction is needed is to determine the origin of the "dimensionful" mass terms in the Higgs' particle equations themselves. At present little if any thought has been given to the origin of these terms. We suggested that, excluding a "deus ex machina" source, the only known way to generate these mass terms in the Higgs' equations is through the separation of equations technique of differential equations. This technique requires additional parameters which can only be the coordinates of extra unknown dimensions. A major example of the generation of mass terms appears in the Schwarzschild solution of General Relativity where a separation constant, often denoted M, appears that has the dimension of [mass]. The extra space-time dimensions required for the Higgs case can only be those of another universe.

This resolution of the origin of the Higgs mass terms also provides an origin for inertial reference frames that replaces Mach's suggestion of the use of "fixed stars" with a sister universe (or set of other universes). The other universe(s) makes inertial reference frames special! When the complete multiverse is introduced then the Flatverse then effectively assumes the role of the "fixed stars" – the Flatverse takes the role of defining inertial reference frames for all its resident universes.

Another major reason for requiring a sister universe, and more generally the multiverse, is the need to implement the Copenhagen interpretation for quantum gravity. The Flatverse (or sister universe) provides the external observer needed for quantum gravity. In addition, as DeWitt[8] and others have emphasized, our universe needs a "clock" for quantum gravity. Again the sister universe and more generally the multiverse provides the required clock

---

[8] DeWitt, B. S., Phys. Rev. **160**, 1113 (1987).

mechanism. Thus the standard Copenhagen interpretation of quantum gravity requires the multiverse. There are also workarounds for these needs but they are generally considered unconvincing.

1.3 Features of Island Universes

In this section we overview features of the island universes within a multiverse. While the internal distribution of matter and energy within a universe can vary there are certain common features of universes that we will discuss here.

First we note that universes are localized agglomerations of mass and energy that are held together by gravity. There structure is determined by Einstein's dynamical equations.

Their space-time region has a complex, four-dimensional metric.[9] While, as we showed in Blaha (2014) other dimensions are possible for a universe, four dimensions is the minimum number of dimensions if one accepts Asynchronous Logic as necessary for a physics that supports coordinated processes that proceed in parallel although separated in space and time.[10] Thus the choice of the number of space-time dimensions is well motivated. The requirement of invariance under *complex* Lorentz transformations in the flat *complex* space-time approximation to locales in galaxies leads to an extended Standard Model with precisely the known internal group symmetries (plus an extra $SU(2) \otimes U(1)$ symmetry that we attribute to Dark Matter). The Standard Model symmetries follow from requiring that complex coordinates must be mapped to real values since rulers and clocks measured real-valued quantities.

Complex General Relativity being founded on Riemannian geometry can also be expected to hold in all universes.

Thus we have a well-founded fundamental theory for all universes: their dimensionality and coordinates, their Standard Model and their General Relativity. Universes expand or contract based on their mass-energy content.

---

[9] Island universes also have the set of four complex coordinates of the Flatverse. The four dimensional complex curved coordinates of a universe are directly related, point by point, to the sixteen dimensional complex flat coordinates of the Flatverse.

[10] See Blaha (2011c) and (2013) as well as Fant (2005) for a description of Asynchronous Logic. The appendices of Blaha (2011c) describe our earlier work on the similarity of particle interactions, finite state machines, and logic.

All of their features are masked since their total energy is zero as are their total internal quantum numbers with one exception: they have a non-null total baryon number.[11] Since baryon number appears to be conserved too great precision we have developed an abelian gauge theory for baryons (Blaha (2014)) that yields (extremely weak) interactions between baryons and thus between universes with non-zero total baryon number. This gauge field implies a conservation law for baryon number. It will be the basis for exiting/entering our universe as well as seeing/navigating in the Flatverse. We discuss it in detail later.

In addition to baryon number, universes may have spin. They can have spin 0 or they can have spin 127/2 since they exist in a sixteen dimensional space. As in the case of elementary particles[12] their spin is not directly connected to their internal motions. Their spin does play a role in the baryonic force interactions of universes. See Blaha (2014) for a detailed discussion.

Lastly, quantum gravity theory, as embodied in DeWitt's paper deriving the Wheeler-DeWitt equation, is applicable to universes. Blaha (2014) describes the important consequences of the *complex* Wheeler-DeWitt equation that yield an understanding of the origin of the Cosmological Constant, of the spatial asymmetry of our universe, and of the web of Galaxies of our universe.

The Wheeler-DeWitt equation must apply to all universes since it is a consequence of Quantum Gravity. Thus other universes of similar age should have features similar to those of our universe: a Cosmological Constant, spatial asymmetry, and a web of Galaxies.

1.4 Large Scale View of the Multiverse

There are two general ways to view the multiverse. We can view it on a large scale with universes appearing as small black spots as in Fig. 1.1. Universes could then be considered to be a form of particle that could move in response to baryonic field force. They could also expand or contract depending on the nature of their interior. In Blaha (2014) we considered the dynamics of universe

---

[11] The only other long range field, the electromagnetic field, is also confined to universes since all universes are infinite in their curvilinear coordinates and no universe has charges at ∞ in their curvilinear coordinates.

[12] Electrons have spin ½ but their spin is not related to their internal structure. Attempts to connect their spin and internal structure (such as Mie's theory) fail by, for example, having the outer electron edge moving faster than the speed of light.

particles. We attributed a mass[13] to universe particles that was not its intrinsic energy (which is zero) but rather an indirect measure of its size based on the spatial area of the universe at a given time:

$$M(t) = \kappa A(t)/8\pi \qquad (1.1)$$

where κ is Boltzmann's constant and $A(t)$ is the area of the universe particle at time t. We then proceeded to develop fermionic-like lagrangian models (neglecting the inner dynamics of the universe particle) such as[14]

$$\mathcal{L} = \bar{\psi}(Y(y))[i\gamma^\mu \partial/\partial y^\mu - e_B\gamma^\mu B_{u_\mu}(Y(y)) - m(t)]\psi(Y(y)) - \tfrac{1}{4} F_{Bu}^{\mu\nu}(Y(y))F_{Bu_{\mu\nu}}(Y(y)) -$$
$$- \tfrac{1}{4} F_u^{\mu\nu}(y)F_{u_{\mu\nu}}(y) \qquad (6.19)$$

We found there were four general types of lagrangian particles: two tachyonic types and two types with complex spatial momentum.

The tachyonic types could be either left-handed or right-handed. The physical meaning of the handedness of these types of universes is an interesting issue. When we consider our universe we see left-handedness in the weak interactions of elementary particles. In addition it appears that organic molecules overwhelmingly favor left-handedness on earth although right-handed molecules exist in outer space and can be created in the laboratory. Right-handed molecules transform into left-handed molecules in watery media through electromagnetic effects.

Why nature favors left-handedness is an open question. It has given rise to speculations that gravitation, especially quantum gravitons, may be left-handed. The European Space Agency's Planck telescope will study polarization effects in the cosmos and may well be able to show that the gravitons starting from the beginning of the universe, and magnified by inflation in the universe's expansion, may be left-handed.

If handedness of gravitation is verified experimentally, then our theory of left-handed/right-handed universe particles would be substantiated. Our

---

[13] This definition follows from Wald's discussion of black holes: Wald, R. M., "The Thermodynamics of Black Holes", *Living Reviews in Relativity* **4** (6): 12119 (2001). We have extended it to universes because a universe, in part, has some qualitative resemblance to a black hole.

[14] From Blaha (2014). This is one of several models. The other models are for faster-than-light (tachyonic) universes and universes with complex spatial momentum.

universe would then be tachyonic and most likely left-handed. It would also imply our universe has spin. Thus we see preliminary suggestions of a particle-like aspect of universes.

If universes have handedness then they also have spin. We developed a half-integer spin type of universe in sixteen dimensions. Universes particles would then have wave functions with 256 components and sixteen 256×256 Dirac matrices. The interpretation of spin states is straight-forward. We suggested that the upper (128) components (64 "spin up" and 64 "spin down") of a universe wave function represent a universe with an excess number of baryons. The lower (128) components lead to anti-universes where there is an excess of anti-baryons.

Since there are 256 possible spin values, using the equation $2s + 1$ = total number of spin values we see that the spin of a fermionic universe particle is $s = 127/2$. The possible universe particle spin values are:

Up spin values: +1/128, +2/128, … , +64/128
Down spin values: -64/128, -63/128, … , -1/128

These associations are analogous to the interpretations of the Dirac electron wave function.

We thus have a particle theory of universes based on a large scale view. We have briefly looked at the odd half integer spin case above. Blaha (2014) has a much more complete description including a description of bosonic universe particles.

Possible evidence has recently appeared for a universe particle dynamics with universes interacting and accelerating due to the fifth force: baryonic gauge fields. Our universe appears to be lopsided: opposite sides of the universe have different temperatures and mass configurations. The lopsidedness could be explained by the acceleration of our universe by a baryonic gauge field force attraction to another antiuniverse – rather like the effect on passengers in an accelerating automobile.

1.5 Small Scale View of the Multiverse Universes

In the preceding section we examined the large scale view of the multiverse and saw universes could be treated as particles. In this section we discuss the small scale view in which the internal structure of universes is examined. These discussions are reminiscent of the development of particle physics. Initially at lower energies (large scale view) protons and neutrons were treated as elementary particles. Subsequently, at much higher energies their internal structure was revealed to consist of interacting quark-partons.

The small scale view of the multiverse is the examination of the internal structure of universes. In Blaha (2014) we saw that this could be done in theory using high energy baryonic gauge field particles that we called *plancktons*. These particle probes from another universe could probe both the classical structure of a universe as well as the quantum structure.

The classical structure of universes is well known and based on General Relativity although there are some questions about the details of General Relativity. The quantum aspects of universes is thought to follow from the Wheeler-DeWitt equation which, itself, is derived from General Relativity suitably quantized. Thus it also is understood in principle.

In Blaha (2014) we generalized the Wheeler-DeWitt equation to complex metrics and found it (and General Relativity) must have additional terms when the Faddeev-Popov mechanism was applied to extract the real metric version of the Wheeler-DeWitt equation and the Einstein lagrangian. The physical reason for restricting the metric to real values is that all space and time measurements with rulers and clocks yield real number values.

Upon finding the extra terms in the Wheeler-DeWitt equation and the Einstein lagrangian we noted new linear features and asymmetries not present in the original equation, and lagrangian. These new features could explain recent new results from astrophysical studies of the universe: a cosmological constant, a web of galaxies, and a spatial asymmetry. Thus we may have a resolution of these outstanding new astrophysical observations.

1.6 The Fifth Force – A Sixteen Dimensional Baryonic Gauge Field

This section[15] describes a fifth force of nature that was presented in Blaha (2014) that straddles the universes of the multiverse and makes it possible for space ships to escape a universe, and also provides navigation capabilities for

---

[15] The material in this section and the following section were first presented in Blaha (2014) which contains significantly more detail.

travels to other universes. This force also provides a mechanism for universe creation and interaction as well as baryon particle interactions.

The primary rationale for the fifth force is the apparent conservation of baryon number. The conservation of baryon number has been repeatedly investigated by experimenters and found to be true to extremely high accuracy. For decades theorists have suggested that a baryon conservation law[16] follows from the existence of an abelian gauge field in a manner much like electric charge conservation follows from the properties of the electromagnetic abelian gauge field.

We therefore assumed in Blaha (2014) that an abelian baryonic gauge field exists that is similar to the electromagnetic field except for features due to its existence in the 16-dimensional Flatverse. This field couples extremely weakly[17] to individual baryons as well as to universe particles due to their non-zero baryon number. We will call the baryonic gauge field particle a *planckton*. Its electromagnetic analogue is the photon.

Plancktons propagate in the Flatverse, both within universes, and exterior to universes. So the planckton field must be defined in 16-dimensional Flatverse coordinates. They will interact with baryons within a universe with Flatverse coordinates mapped to the curved coordinates in the universe.

Since a planckton field in 16-dimensional conventional coordinates would lead to divergences we use quantum coordinates:[18]

$$Y^i(y) = y^i + i\, Y_u^i(y)/M_u^8 \qquad (1.2)$$

with quantum coordinate derivatives defined by

$$\partial_i = \partial/\partial Y^i(y) = \partial/\partial(y^i - Y_u^i(y)/M_u^8) \qquad (1.3)$$

to obtain a completely finite theory of planckton interactions with elementary particles and universe particles.

---

[16] See Gell-Mann, M. and Levy, M. *Nuovo Cimento* 16, 705 (1960) for a proof and Sakurai (1964) for a discussion of the relation of the baryonic gauge field to gravity experimentally.
[17] Compared to gravity.
[18] See Blaha (2005a) for a discussion of this new method to eliminate infinities in quantum field theory calculations.

Plancktons and the $Y_u^i(y)$ field of quantum coordinates are the only fields in the space between universes in the Flatverse. Since the mass-energy and charge of universes is zero, gravitation and Standard Model fields are zero in the space between universes.[19] It is reasonable to assume that the vacuum between universes does have fermion or boson seas for Standard Model particles.

1.6.1 Planckton Second Quantization

The second quantization of the free planckton field $B_u^i(y)$ is similar to the second quantization of the electromagnetic field, and also of the quantum part of the Flatverse quantum coordinates $Y_u^i(y)$. The purpose and role of these fields is quite different: the planckton field generates an interaction between baryons while the $Y_u^i(y)$ field serves as the quantum part of 4-dimensional quantum coordinates giving us a finite quantum field theory of The New Standard Model and gravitation as well as a finite Big Bang for our universe.

We begin by noting that Flatverse quantum coordinates are defined by eqns. 1.2 and 1.3 above. The lagrangian density terms for the free $B_u^i(Y(y))$ fields is

$$\mathscr{L}_{Bu} = -\tfrac{1}{4}\, F_{Bu}{}^{\mu\nu}(Y(y))F_{Bu_{\mu\nu}}(Y(y)) \tag{1.4}$$

with Y(y) given by eq. 1.2. The lagrangian is

$$L_{Bu} = \int d^{15}y\, \mathscr{L}_{Bu}(Y(y)) \tag{1.5}$$

with

$$F_{Bu_{\mu\nu}} = \partial B_{u_\mu}(Y(y))/\partial Y^\nu(y) - \partial B_{u_\nu}(Y(y))/\partial Y^\mu(y) \tag{1.6}$$

where the values of μ and ν range from 1 to 16 in this section.

The equal time commutation relations, derived in the usual way, are:

$$[B_u{}^\mu(Y(\mathbf{y}, y^0)), B_u{}^\nu(Y(\mathbf{y}', y^0))] = [\pi_u{}^\mu(Y(\mathbf{y}, y^0)), \pi_u{}^\nu(Y(\mathbf{y}', y^0))] = 0 \tag{1.7}$$

$$[\pi_{uj}(Y(\mathbf{y}, y^0)), B_{uk}(Y(\mathbf{y}', y^0))] = -i\,\delta^{15tr}{}_{jk}(Y(\mathbf{y},0) - Y(\mathbf{y}',0)) \tag{1.8}$$

where

---

[19] The vacuum energy of the baryonic field and the $Y_u^i(y)$ fields being uniform throughout the Flatverse do not exert forces or cause gravitational effects except possibly through baryonic Casimir forces between universes.

$$\pi_u{}^k = \partial \mathscr{L}_u (B_u(Y(y)))/\partial B_{uk}'(Y(y)) \tag{1.9}$$

$$\pi_u{}^0 = 0 \tag{1.10}$$

and

$$\delta^{tr}{}_{jk}(\mathbf{y} - \mathbf{y}') = \int d^{15}k \; e^{i \, \mathbf{k} \bullet (Y(y,0) - Y(y',0))} \, (\delta_{jk} - k_j k_k/\mathbf{k}^2)/(2\pi)^{15} \tag{1.11}$$

$$B_{uk}'(Y(y)) = \partial B_{uk}(Y(y))/\partial y^{16} \tag{1.12}$$

for j, k = 1, 2, ... , 15.

If we choose the Coulomb gauge for $B_{uk}(Y(y))$:

$$B_u{}^{16}(Y(y)) = 0$$

$$\partial B_u{}^j(Y(y))/\partial Y^j(y) = 0$$

for j = 1, 2, ... , 15 then fourteen degrees of freedom (polarizations) are present in the vector potential.[20] The Fourier expansion of the vector potential $B_u{}^i(Y(y))$ is:

$$B_u{}^i(Y(y)) = \int d^{15}k \; N_{0B}(k) \sum_{\lambda=1}^{14} \varepsilon^i(k, \lambda)[a_B(k,\lambda) :e^{-ik \cdot Y(y)}: + \, a_B{}^\dagger(k,\lambda) :e^{ik \cdot Y(y)}:] \tag{1.13}$$

for i = 1, ... , 15 where

$$N_{0B}(k) = [(2\pi)^{15} 2\omega_k]^{-\frac{1}{2}} \tag{1.14}$$

and (since the field is massless)

$$k^{16} = \omega_k = (\mathbf{k}^2)^{\frac{1}{2}} \tag{1.15}$$

where k^{16} is the energy, and where the $\varepsilon^i(k, \lambda)$ are the polarization unit vectors for λ = 1, ... , 14 and $k^\mu k_\mu = k^{16\,2} - \mathbf{k}^2 = 0$.

The commutation relations of the Fourier coefficient operators are:

$$[a_B(k,\lambda), a_B{}^\dagger(k',\lambda')] = \delta_{\lambda\lambda'} \delta^{15}(\mathbf{k} - \mathbf{k}') \tag{1.16}$$

$$[a_B{}^\dagger(k,\lambda), a_B{}^\dagger(k',\lambda')] = [a_B(k,\lambda), a_B(k',\lambda')] = 0 \tag{1.17}$$

and the polarization vectors satisfy

$$\sum_{\lambda=1}^{14} \varepsilon_i(k, \lambda)\varepsilon_j(k, \lambda) = (\delta_{ij} - k_i k_j/\mathbf{k}^2) \tag{1.18}$$

---

[20] Note we use the Coulomb gauge for Y(y) also.

The $B_u{}^\mu$ Feynman propagator is

$$iD_F{}^{trTT}(y_1 - y_2)_{jk} = <0 | T(B_{uj}(Y(y_1))B_{uk}(Y(y_2))) | 0> \qquad (1.19)$$

$$= - ig_{jk} \int \frac{d^{16}k \, e^{-ik\cdot(y_1 - y_2)} R(\mathbf{k}, y_1 - y_2)}{(2\pi)^{16} (k^2 + i\varepsilon)} \qquad (1.20)$$

where g_{jk} is the 16-dimensional Lorentz metric and where $R(\mathbf{k}, y_1 - y_2)$ is given by

$$R(\mathbf{k}, y_1 - y_2) = \exp[- k^i k^j \Delta_{Tij}(y_1 - y_2)/M_u{}^{16}] \qquad (1.21)$$
$$= \exp\{ -k^2[A(v) + B(v)\cos^2\theta] / [(2\pi)^{14}M_u{}^4 z^2] \}$$

with

$$z^\mu = y_1{}^\mu - y_2{}^\mu$$
$$z = |\mathbf{z}| = |\mathbf{y_1} - \mathbf{y_2}|$$
$$k = |\mathbf{k}|$$
$$v = |z^0|/z$$
$$A(v) = (1 - v^2)^{-1} + .5v \, \ln[(v - 1)/(v + 1)]$$
$$B(v) = v^2(1 - v^2)^{-1} - 1.5v \, \ln[(v - 1)/(v + 1)]$$
$$\mathbf{k \cdot z} = kz \cos\theta$$

and $|\mathbf{k}|$ denoting the length of a spatial 15-vector $\mathbf{k}$ while $|z^0|$ is the absolute value of $z^0 \equiv z^{16}$.

As eq. 1.21 indicates, the Gaussian damping factor $R(k, z)$ for all large spatial momentum k^j is the same for both the positive and negative frequency parts of the (two-tier) B_u Feynman propagator. We are assuming the spatial momentum is real-valued in this discussion. It is also important to note that $R(k, z)$ does not depend on $k^0 = k^{16}$ (in the B_u and Y_u Coulomb gauges) and thus the integration over k^0 proceeds in the usual way to produce time-ordered positive and negative frequency parts.

The Gaussian exponential factor in *all* spatial coordinates causes the Feynman propagator to be finite and, together with the Gaussian factor in universe particle propagators, causes all perturbation theory calculations when interactions are introduced to be finite as we have seen earlier in The New Standard Model.

For small momentum much less than M_u then $R(\mathbf{k}, y_1 - y_2) \to 1$ and the Feynman propagator is the "normal" propagator of conventional 16-dimensional quantum field theory. For large momentum the corresponding potential approaches r^{13} in contrast to the electromagnetic Coulomb potential r^{-1}. The B_u potential is highly non-singular at large energies.

1.6.2 Planckton Interactions between Universe Particles and Individual Baryons

In this section we will develop an interacting theory of universe particles and plancktons from the lagrangian terms of universe particles, plancktons and quantum coordinates. We will only consider the case of Dirac type universe particles since the other cases differ from it only in details.

$$\mathcal{L} = \bar{\Psi}(Y(y))[i\gamma^\mu \partial/\partial y^\mu - e_B\gamma^\mu B_{u\mu}(Y(y)) - m(t)]\psi(Y(y)) - \tfrac{1}{4} F_{Bu}{}^{\mu\nu}(Y(y))F_{Bu\mu\nu}(Y(y)) -$$
$$- \tfrac{1}{4} F_u{}^{\mu\nu}(y)F_{u\mu\nu}(y) \tag{1.22}$$

where $\mu, \nu = 1, 2, \ldots, 16$ and where

$$\bar{\Psi} = \psi^\dagger \gamma^{16}$$
$$F_{Bu\mu\nu} = \partial B_{u\mu}(Y(y))/\partial Y^\nu(y) - \partial B_{u\nu}(Y(y))/\partial Y^\mu(y)$$
$$F_{u\mu\nu} = \partial Y_\mu/\partial y^\nu - \partial Y_\nu/\partial y^\mu$$
$$Y^i(y) = y^i + i\, Y_u^i(y)/M_u^8$$
$$e_B = e_{B0}/M_u^6$$

with e_{B0} a dimensionless coupling constant, and with μ and ν ranging from 1 through 16.

The lagrangian is

$$L = \int d^{15}y\, \mathcal{L} \tag{1.23}$$

Note the dimensions of the fields differ in the 16 dimensional space:

$$Y^\mu \sim [\text{mass}]^7$$
$$B_{u\mu} \sim [\text{mass}]^7$$

$$\psi \sim [\text{mass}]^{15/2}$$

as can be seen from the above lagrangian as well as earlier equations. Note also that the mass and thus the size of universe particles is time dependent. They can expand or contract with time depending on their internal characteristics (gravitation and effects of elementary particle interactions) which are not embodied in this lagrangian. As a result this theory, incomplete as it is, does not conserve energy unless $m(t)$ is constant.

The lagrangian generates the baryonic interactions of universe particles using Two Tier quantum coordinates which prevent infinities in perturbation theory calculations.

The interaction of baryon elementary particles with the baryonic field requires terms in The New Standard Model specifying the baryon field interaction baryons with the form

$$e_B \gamma^\mu B_{u_\mu}(Y(y))$$

The following sections describe some of the physically significant interactions that the lagrangian implies.

1.6.3 Universe Particle – Planckton Interaction Example

This type of interaction is quite similar to two-tier electromagnetic interactions except that universe particles have time-dependent masses, and that the space is 16-dimensional.

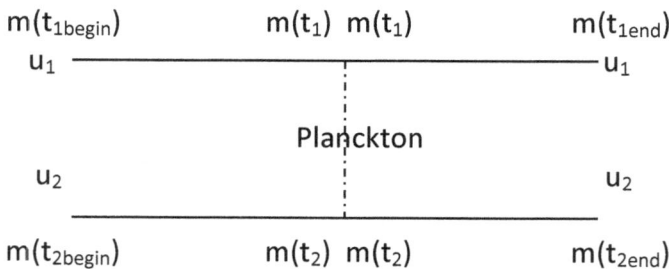

Figure 1.2. A Feynman diagram illustrating the continuity of a universe particle mass through a Planckton interaction.

The interactions have a new aspect due to the time dependence of the universe particle masses. This feature is illustrated by Fig. 1.2: the mass of a universe particle after a baryonic interaction vertex is the same as it was before the interaction assuming the point-like interaction specified in the lagrangian.

The reader may verify this by writing the perturbation theory equivalent. A universe particle vertex corresponds to

$$iS_F^{TT}(y_1, y_2)\gamma^\mu iS_F^{TT}(y_2, y_3) \tag{1.24}$$

Note the universe particle mass is the same on either side of the interaction vertex.

1.6.4 Types of Baryonic Interactions

We have developed a planckton field theory that gives interactions between baryons. This theory is applicable to universe-universe interactions. It also yields baryon particle – baryon particle interactions as well as baryon particle – universe particle interactions.

It is possible for a planckton to be emitted in one universe and interact with a baryon elementary particle in another universe. This type of "probe" must be a high energy probe just as a photon probe of the internal structure of a nucleon[21] must be a high energy photon to bring out the nucleon's internal structure (parton model).

In the next section we will discuss planckton probes of other universes.

1.6.5 Planckton Probes

Plancktons can be generated in one universe and be used to probe the baryon distribution of another universe. Since the planckton propagator is expressed in Flatverse coordinates the baryon distribution in the target universe will be a distribution in Flatverse coordinates. Flatverse coordinates can be expressed in terms of the curved space-time coordinates of a universe x^μ Universe and Flatverse coordinates are related by

$$y_i = f_i(x) \tag{1.25}$$

---

[21] Deep inelastic electron-nucleon scattering.

with the inverted relation having the form

$$x^\mu = f^{-1\mu}(y) \tag{1.26}$$

The inverted relation eq. 1.26 is not 1:1 since x^μ is 4-dimensional and y is a 16-dimensional vector. The universe coordinates x^μ are each individually determined up to a subspace. One might be concerned about this situation but the determination of the distribution in Flatverse coordinates gives a more direct picture not convoluted by the curvature of the target universe.

The detailed probe of a target universe requires high energy plancktons. The similarity of this procedure to deep inelastic electro-nucleon scattering is obvious to the high energy physicist. But in doing a planckton probe experiment one obtains a picture of a different universe – something that is not possible to do with electromagnetic or graviton probes.

1.6.6 Full Lagrangian of Universe Particles and Plancktons

The above development of the theory of universe particles does not fully describe universe particles since it neglects the internal structure of a universe particle. The internal structure of a universe particle is primarily determined by gravitation, electromagnetic effects and nuclear physics.

Consequently the full lagrangian of a universe particle has the form

$$\mathcal{L}_{tot} = \mathcal{L}_{internal} + \mathcal{L} \tag{1.27}$$

where $\mathcal{L}$ is determined by eq. 1.22. As a result the complete quantum wave function of a universe particle has the form

$$\psi_{tot} = \psi_{internal}(Y)\psi_{ext}(Y) \tag{1.28}$$

where $\psi_{internal}(Y)$ is the internal wave function and $\psi_{ext}(Y)$ is determined from the lagrangian eq. 1.22. It seems reasonable to have a separable equation as above, except when universes collide. In that situation a perturbative mixing of the universes and their wave functions applies and it may be possible to calculate the collision output universes by introducing a further interaction between the internal and external aspects of the universe particles.

1.7 The Origin of Universes

The origin of the universes of the multiverse is a difficult question to answer. The simplest general answer postulates all universes are generated ultimately from vacuum fluctuations of the baryonic field. Subsequent to their creation they may combine or fission into new generations of universes IF the time for the recombination of universes into the vacuum is sufficiently long. This scenario resolves the problem of the original creation of universes within an initially empty multiverse. In this section we will examine creation from vacuum fluctuations, universe collisions, the coalescence of universes and the fission of universes.

1.7.1 Creation of Universes through Baryonic Gauge Field Fluctuations

One of the most exciting questions in Cosmology is the origin of our universe. The conventional view is that it originated in a Big Bang from an infinitesimal point in space. The source of the Big Bang and the prior state of the Cosmos, if there was one, is the subject of much speculation. Based on the particle interpretation of the Wheeler-DeWitt equation, the possibility of a baryonic force strongly supported by conservation of baryon number, and the multiverse concept it is reasonable to consider the possibility that our universe originated in a vacuum fluctuation.

Our formulation of universe particle theory provides for the generation of a universe particle and anti-particle as a vacuum fluctuation. We view a universe particle as having a substantial excess of baryons, N, as we see in our universe. Its anti-universe at the time of creation (the Big Bang point) is its "mirror image" having the same number of anti-baryons (baryon number: −N) so that baryon number is conserved by the fluctuation event.

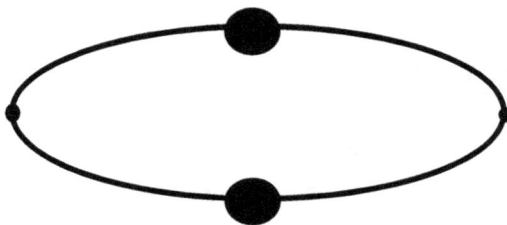

Figure 1.3. Generation of a universe – anti-universe pair as a vacuum fluctuation.

The small value of the coupling constant could lead to an extremely long lifetime for the universes generated by a fluctuation. Thus the 13.7 billion year life of our universe is not unreasonable. Its lifetime can be extremely long. The probability of the creation of universes by vacuum fluctuations should be correspondingly small.

1.7.2 When Universes Collide: Coalescence of Universes

Universes moving in the Flatverse can collide through chance, or due to the planckton field which causes universes with excess baryons to attract universes with excess anti-baryons.

When universes collide several possibilities present themselves:

1. They can graze each other distorting each other's shape and internal baryon distribution through the baryonic force while maintain their individual identity.
2. They can intermix with both the baryonic and gravitational forces causing a redistribution of their masses. They may separate afterwards or may coalesce into a single universe. One result of this may be lopsided universes. *Our universe appears to be lopsided.* Some cosmologists believe this is due to a near collision of our universe with another shortly after the Big Bang.

1.7.3 Fission of Universes

Under certain circumstances the distribution of matter in the universe may lead to the fission of a universe into two separate universes. Our model lagrangian supports this possibility for universe particles. The detailed mechanism of the fission process is not specified by the model.

1.7.4 Fission of "Normal" universes

The fission of normal (non-tachyon) universe particles in our universe particle model is depicted in the Feynman diagram in Fig. 1.4. The sum of the masses of the output universe particles is usually less than the original universe particle mass. However if the fission takes a long time and the masses are time dependent then the combined masses of the produced universe particles may exceed the original universe's mass.

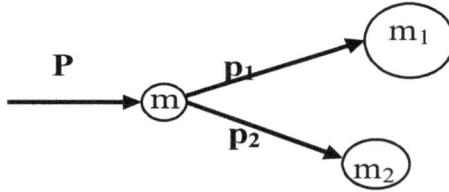

Figure 1.4. Fission of a universe particle into two universe particles.

1.7.5 Tachyon Universe Particle Fission to More Massive Universe Particles

In Blaha (2007a) we showed that a tachyonic (faster than light) particle could fission into particles of larger mass. In this section we will show that a tachyonic universe particle may fission into more massive universe particles. This phenomenon is of particular interest because it enables tachyonic universes to spawn in a new novel way not previously considered in discussions of the origin of universes.

The lagrangian for a tachyonic universe particle is

$$\mathcal{L}_{||} = \psi_T^{~S}(Y(y))[\gamma^\mu \partial/\partial y^\mu - e_B \gamma^\mu B_{u\mu}(Y(y)) - m(t)]\psi(Y(y)) - \tfrac{1}{4}\,F_{Bu}^{~\mu\nu}(Y(y))F_{Bu\mu\nu}(Y(y)) -$$
$$- \tfrac{1}{4}\,F_u^{~\mu\nu}(y)F_{u\mu\nu}(y) \tag{1.29}$$

For simplicity we will assume m(t) is constant.

When a particle or a universe particle fissions (decays) one normally expects that the masses of the particles or universe particles produced by the decay to be smaller than the mass of the original particle or nucleus. In the case of tachyonic (faster-than-light) elementary particles or universe particles a much different possibility is present: a tachyon can decay into heavier tachyons. We will consider the specific case of a tachyon universe particle decaying into two universe particles whose total mass is greater than the original. (See Fig. 1.5.)

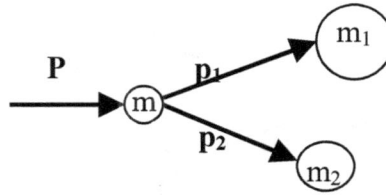

Figure 1.5. Two universe particle decay of a tachyon universe particle.

We will assume the initial tachyon universe particle has zero energy (p^{16} = 0) and thus the tachyons universe particles emerging from the decay also have total universe particle energy zero. The analysis is based on conservation of total universe energy and momentum in the Multiverse. The below discussion applies to 16-dimensional space with 15-dimensional spatial coordinates.

Momentum conservation implies

$$P = p_1 + p_2 \qquad (1.30)$$

Since all energies are zero

$$(cP)^2 = (c\mathbf{P})^2 = m^2$$
$$(cp_1)^2 = (c\mathbf{p_1})^2 = m_1^2 \qquad (1.31)$$
$$(cp_2)^2 = (c\mathbf{p_2})^2 = m_2^2$$

where $P = |\mathbf{P}|$, $p_1 = |\mathbf{p_1}|$, and $p_2 = |\mathbf{p_2}|$. If we now square eq. 1.20 and then use eqns. 1.31 we obtain

$$m^2 = m_1^2 + m_2^2 + 2m_1m_2 \cos\theta \qquad (1.32)$$

where θ is the angle between the emerging universe particles momenta $\mathbf{p_1}$ and $\mathbf{p_2}$. Eq. 1.32 has a number of interesting cases:

Case $\theta = 0$:

$$m = m_1 + m_2$$

The masses of the outgoing universe particles sum to the mass of the original tachyon universe particle.

Case $\theta = \pi/2$:

$$m^2 = m_1^2 + m_2^2$$

The masses of each outgoing universe particle tachyon is less than the mass of the original tachyon universe particle.

Case $\theta = \pi$:

$$m^2 = (m_1 - m_2)^2$$

In this case either $m_1 > m$ or $m_2 > m$. Thus one of the outgoing tachyon universe particles has a greater mass than the original tachyon universe particle. Mass is effectively created from the spatial momentum of the initial universe particle. This process is the inverse of normal particle and universe particle fission where the sum of the outgoing masses is always less than the original particle's mass and the difference is mass converted into energy in the form of additional photons.

This last case, where one of the outgoing universe particles is more massive than the original universe particle, is not just for $\theta = \pi$. Since

$$\cos \theta = (m^2 - m_1^2 - m_2^2)/(2m_1m_2)$$

we see that the sum of the outgoing universe particle masses is always greater than the original tachyon universe particle mass (except when $\theta = 0$) since

$$\cos \theta = 1 + [m^2 - (m_1 + m_2)^2]/(2m_1m_2) \leq 1$$

and thus

$$[m^2 - (m_1 + m_2)^2]/(2m_1m_2) \leq 0$$

Note $m = m_1 + m_2$ only if $\theta = 0$.

Since we can transform the above discussion to the case of universe particle tachyons having non-zero multiverse energy using an ordinary 16-dimensional Lorentz transformation the discussion in this subsection is general.

We therefore conclude that when a tachyon universe particle decays into two tachyon universe particles the sum of the masses of the produced tachyon universe particles is greater than the mass of the original tachyon universe particle except if the angle between the momenta of the produced tachyon universe particles is zero. In that case the sum of the masses of the produced tachyon equals the mass of the original tachyon universe particle and the produced universe particles overlap.

1.8 Multiple Multiverses?

We have assumed that there is one multiverse. However if more dimensions exist – in particular an additional sixteen more complex dimensions then there is no reason why more multiverses would not exist. The possibility of a countable infinity of multiverses is not excluded.

Other multiverses would not be accessible unless some interaction existed that straddled the multiverses. Such an interaction would have to be super weak to avoid a significant interplay between the multiverses – perhaps even a mixing of multiverses.

A possible inter-multiverse force would have to be a function of all the coordinates of all multiverses. Based on an analogy with the multi-universe baryonic force the form of the inter-multiverse force could be an abelian gauge field. The inter-multiverse force charge would be perhaps the number of universes in a multiverse minus the number of anti-universes, N. This quantum number may or may not be conserved. The inter-multiverse force may or may not have an associated conservation law similar to the conservation law for electric charge.

The simplest multiple multiverse cosmos would exist in a thirty two dimension Euclidean space with each multiverse in a sixteen subspace.[22] The multiverses could not overlap or they would then constitute one multiverse. Each multiverse would occupy all, or part, of a sixteen dimensional subspace. The multiverses in this space would move in response to dynamical equations similar to elementary particle dynamical equations. The existence of a multiple multiverse force could only be detected by the acceleration of our multiverse. This possibility lies in the far distant future at best.

---

[22] It would be possible to "shoehorn" two multiverses into a seventeen dimensional space with each multiverse occupying a sixteen dimensional subspace part with no overlap of multiverses.

2 Why Explore the Multiverse?

Expansion and exploration have been occupations of Man since well before the beginnings of recorded history – indeed since before Man emerged as a species. An examination of history shows that exploration has been of great importance for the growth of Man as a species. Perhaps the most important recent example is the discovery of the Americas which provided not only land for farming and mining but also many new plant varieties (such as corn and the potato) that have greatly enriched the entire world.

Today Man is beginning the exploration of the solar system.[23] In perhaps a hundred years Man will have the technology and the resources to travel to the stars. While multi-generation starships for slower than light travel are currently the most popular approach, it is our hope – based on our Extended Standard Model theory of elementary particles – that much faster than light starships will be built and open the stars for colonization and commerce: The stars – the new America!

After Man has explored and colonized the stars of our galaxy, and then of the other galaxies of our universe, the challenge of exploring the multiverse will present itself. This challenge will not appear soon; but Man will eventually think this universe is too small for his ambitions. Then the multiverse with its multitude of universes will offer the opportunity to see, and learn from, all the Cosmos. If we must make a guess as to when this challenge will be realized, then it would seem that it will likely be in fifty thousand years, plus or minus, barring scientific developments of an extraordinary nature.

One might ask why we should consider such a remote future which seems to betoken science fiction more than science. The only significant answer to this query is that the knowledge of this distant goal will help guide Man's

---

[23] We have discussed the best approaches to space travel in earlier books: Blaha (2009a) (2009b), (2011b), (2013a). These books have played a significant role in promoting new space initiatives – especially the NASA starship program. The new US Navy heavy electromagnetic guns and rail guns, if extended, can become space guns for the cheap transport of materials from the earth's surface to earth orbit at **enormously lower cost** making space easily accessible to Man. We briefly indicate these possibilities, especially faster than light, quark-gluon ion drive starships in appendix A.

effort towards its eventual achievement. There is a path within our view. It may not succeed. But it is worthy to consider, and to investigate, over the next millennia. We hope that the challenge of that path will have continuing appeal to subsequent generations of scientists and technologists.

As Arnold Toynbee concluded after his monumental study of civilizations, Man can only progress by accepting new challenges and overcoming them.

Simply put: species either grow or stagnate.

2.1 Benefits for Man in Exploring the Multiverse

Most of the semi-intelligent (intelligent?) species on earth that we have encountered have found a niche in Nature which they can dominate. Whales and elephants dominate their niches by their great size. Primates dominate their niches by intelligence and agility as do dolphins and porpoises. The consequences of a too successful triumph in a niche is stagnation. Whales, for example, have changed very little in the past three hundred million years. Primates have changed because their dominance was not overwhelming and they were confronted with challenges to their survival which they overcame by changing.

Man has achieved dominance over the earth One can foresee a possible future with perhaps new technology but with little change in Man as a species. In fact the great success of medicine has led to a population that appears to be slowly declining physically and mentally on average. Darwinian natural selection has been severely lessened by our humane efforts to cure people, prolong lives, and minister to the handicapped. This may sound callous but growth comes through struggle and the success of the fittest.

The venture into space is an effort to address the challenge of an overcrowded and environmentally degraded earth. One hopes that this effort, if pursued with vigor, will make the planets and moons of the solar system the homes of new civilizations. As in previous history on earth, new civilizations on new "soil" lead the way in Man's progress into the future. America is the clear example of this phenomena but one sees it also in the earlier history of Asia and Europe as old civilizations decay and are "replaced" with new civilizations on the frontiers.

The alternative to the growth of civilizations through expansion into new "land" is a petrified global civilization with a veneer of progress but with a continuation of the current social conditions in a progressively declining planet.[24] The efforts to clean the environment, and improve the earth, in an effectively closed system, such as the earth[25] is, cannot succeed according to the thermodynamic law that entropy always increases in a closed system. An increase in the entropy of the earth is equivalent to a decline in its environment and society.

These thoughts lead us to consider the distant future, where an analogous situation on a larger scale will exist, when Man, having "conquered" the universe, must accept the challenge of a venture into the other universes of the multiverse or become a petrified civilization and a declining species.

What benefits can we expect from exploring the universes of the multiverse? Some of the benefits would be:

1. Encounters with alien civilizations and life that could benefit Man intellectually and technologically just as the Byzantines brought knowledge and civilization to Western Europe leading to a rebirth culminating in the Renaissance.

2. Chemistry and solid state physics are broad areas of study which seem to continually open new possibilities. Thus it is possible that exotic materials and their applications could be found in other universes to the enrichment of Man.

3. It is clear from Blaha (2014) that our universe is subject to dynamic forces in the multiverse. Collisions of universes are possible as well as other calamities. These potential disasters may be much closer in time than the ultimate collapse of the universe, which is often a subject of scientific discussion and news articles. The ability to move part or all of Man into another universe may be a solution to these concerns.[26]

---

[24] No greater example of the degradation of the earth exists than the large amounts of pollution in the southern Indian Ocean recently seen in the search for a missing Boeing 777 jet – far from the industrialized regions of the world.

[25] The earth is not a completely closed system because it receives energy from the sun. However the energy obtained from the sun will not save earth's environment from degradation as the poisoning of the oceans with mercury from gold mining vividly demonstrates.

[26] Professor Hawking has repeatedly made similar, but less grand, proposals to "save" Man.

4. It may be possible to develop new senses, or extended senses, by multiverse exploration from contact with new species or due to new physical environments in other universes.

5. The successful development of colonies and subsequently human civilizations in other universes would again give man the challenge of a new frontier that could further human progress.

6. Additional unforeseen benefits will be likely. We have seen the technology spinoffs from the space program. The spinoffs from multiverse exploration should be of a much larger scale.

So we suggest exploration in the multiverse has multifold advantages for the advancement of Man so that Man, as a species, will not have existed in vain – little better than ants and bees. We should seek to encompass all the Cosmos to extend our understanding to the entirety of mind and matter. This effort, if successful, will make Man a magnificent new species.

The Anglican Book of Common Prayer states "Death is the Great Victory." We should like to amend it to "Universal Knowledge is the Great Victory" that justifies Man as a species.

2.2 Scientific Benefits of Exploring the Multiverse

If we explore the universes of the multiverse it is quite possible that we will find strange new universes with a different physics and/or a different topology. These potential discoveries would include:

1. Universes with more than four dimensions.

2. Universes with different physical laws.

3. Universes which support extensions of the human senses.

4. Universes with different values for coupling constants and/or masses that might help explain their origin and deepen our knowledge of reality.

5. Universes with a different topology that might sharpen our understanding of General Relativity.

6. The capability to perform comparative studies of universes that will deepen our understanding of space and time. At the moment we are limited to one specimen of the set of universes. Comparative studies of universes would make the study of space and time an "experimental" science.

7. Universes with unique new materials that would substantially improve our technology.

8. The development of a uniship that can enter and traverse the multiverse would itself generate major advances in our technology (spinoffs) just as the space exploration program has done in the 20th century.

Thus we see that exploring the multiverse and its universes has broad implications for the far future of Man.

3 Escaping Our Universe – A Neutron Star Slingshot

It is no small matter to escape from the confines of our universe. Euclid's proof that space has three dimensions illustrates the dilemma. Briefly summarized, it states: move in a straight line in any direction; then make a right angle turn and move in another straight line direction; then again make a right angle turn in a direction not parallel to the first direction of movement. Then notice that there is no further right angle turn that does not parallel any of the previous directions of movement. Thus three dimensions according to Euclid. Seemingly inescapable!

In this chapter we propose to escape the confines of our universe using neutron stars to slingshot a uniship into the Flatverse using the neutron stars baryonic field. This mechanism requires a carefully managed uniship trajectory around a neutron star that comes close to the star but can avoid the problems of extremely high gravity, magnetic field and radiation. We will consider these issues in this chapter leaving a discussion of alternate uniship escape mechanisms to chapter 4.

This discussion as well as that of subsequent chapters assumes the existence of an abelian, baryonic gauge field pervading the Flatverse and universes. This field is 16-dimensional and depends on baryon charge. Otherwise it is very similar to the electromagnetic field, which depends on electric charge. If the baryonic field does not exist (It is an open question experimentally at present.), then our approach fails. However the experimentally confirmed conservation of baryon number to very great accuracy strongly indicates that the baryonic field (the "fifth force") does exist because this gauge field would imply baryon number conservation just as the conservation of electric charge is implied by the electromagnetic gauge field. The author is thus confident that this very weak interacting field (compared to all other interactions) will be found and eventually open the door to the multiverse.

Secondly, we must assume uniships can be built that can achieved speeds much, much faster than the speed of light. In our universe we need starships able to eventually reach speeds up to tens of thousands of times the speed of light so that we can travel to distant galaxies in reasonably short times. The distances between galaxies in our universe are typically of the order of millions of light years.

For travel in the multiverse we would need uniships capable of speeds of the order of millions of times the speed of light to successfully travel between universes. We anticipate the distance between universes is of the order of trillions of light years. A saving factor could be a prevalence of wormhole "shortcuts" that uniships could travel through to substantially shorten travel distances.

3.1 How Can We Evade Euclid's Proof of Three Dimensions and Enter Other Dimensions?

We will now consider a mechanism to escape from a lower dimensional space to a higher dimensional space containing the lower dimensional space as a subspace. We will see that the key factor in this escape mechanism is *a force in the higher dimensional space* that boosts an object in the lower dimensional space into dimensions in the higher dimensional space.

A simple example illustrating this mechanism is a two dimensional flat space (often called *Flatland*) within three dimensional space. We will take the Flatland to be the x-y plane and the z-axis to be the other dimension of the three dimensional space.

Suppose a charged particle is moving with speed v in the positive x direction. Suppose further that a constant magnetic force B points in the positive y direction. Then due to the magnetic vector force law for an electric charge q

$$\mathbf{F} = (q/c)\,\mathbf{v} \times \mathbf{B}$$

the force on the charge will be in the positive z direction – popping the charge out of Flatland into the full three dimensional space. Note that the magnetic force is in the Flatland but the 3-dimensional form of the force law causes the force on the charge to propel it from the Flatland into three dimensional space.[27]

---

[27] We assume the escape from the two dimensional subspace is not impeded by a "horizon." We note that this possible impeding horizon does not exist in reality in this ordinary electromagnetism example.

3.2 Escaping Our Universe using the Baryonic Gauge Field

The preceding example illustrates the mechanism for particles and uniships to escape from our four dimensional universe into the sixteen dimensional Flatverse. The force field that will generate the escape is the baryonic gauge field force (the "fifth" force) described in chapter 1 and again in more detail in chapter 5. It can be used to "pop" baryons and uniships (which are primarily composed of baryons) into the Flatverse.

We shall describe various other methods to provide escape from our universe in chapter 4.

3.3 The Neutron Star Slingshot for Uniships

Assuming very fast uniships we can imagine a uniship going at a speed of perhaps half the speed of light almost directly at a neutron star passing close to it in trajectory "bent" by gravity in our three dimensional space and "bent" by the baryonic force into the multiverse – thus breaking it out of the universe into the Flatverse[28] of the multiverse. (Fig. 3.1) This slingshot into the multiverse has important issues that require a carefully (almost certainly computer) managed trajectory. We discuss the major issues that constrain the trajectory in the following subsections.

Figure 3.1. Depiction of slingshot of uniship around a neutron star. The solid line part is the uniship still within our universe. The dashed line part is the uniship trajectory after the baryonic force has bent it out of our universe into the multiverse.

---

[28] The Flatverse is the entire multiverse including universes and the sea between the island universes.

3.3.1 Some Neutron Star Features

Prior to discussing uniship trajectory issues we will review some of the important features of neutron stars. Neutron stars are stars that have collapsed under the force of gravity from a size of 10+ solar masses to a diameter of the order of 10 km after expending their nuclear fuel. They are composed primarily of neutrons although they may have an outer layer of electrons and ions, and an inner core that might constitute a quark-gluon plasma.

The mass of a neutron star lies between 1.44 solar masses (the Chandrasekhar limit) and about 3 solar masses (beyond which it collapses to a black hole).[29] The very small radius of a neutron star and its great relative mass lead to an escape velocity of the order of c/3 (one-third the speed of light).

Neutron stars typically rotate at a rate of 2-3 revolutions per second to a rate of up to 700 revolutions per second.

There are a great many neutron stars in our galaxy. It is estimated that there are of the order of 10^8 neutron stars in our galaxy. The vast number of comparably sized galaxies in our universe show that the number of neutron stars is truly large and thus the sling shot mechanism can be executed from "almost anywhere" in any galaxy in the universe.

3.3.2 Problems and Solution for Close Approach to Neutron Stars

The major problems of a close approach of perhaps 100,000 km to a neutron star are the large gravity of the neutron star, its strong tidal gravitation effects (stresses) on the uniship's structure, its strong magnetic field, and its emission of large amount of primarily x-ray radiation. These properties of a neutron star neighborhood would appear to significantly affect the structural integrity of the uniship, and, more importantly, seriously impact on the safety and life of its human crew.

Fortunately there is a saving grace in this physical environment. A uniship approaching the neutron star could resolve these issues by traveling at extremely high speed[30] so that the time spent in the "danger" zone of the neutron star would be "infinitesimal" reducing neutron star effects to a point much lower than the time interval needed for significant damage to occur.

---

[29] Black holes have baryon number zero and thus do not generate a baryonic field.

[30] The starship would approach at perhaps c/2 and acquire an additional speed due to gravity of c/3. Combining these speeds using the rules of special relativity for the addition of velocities yields an approach speed of 0.7c near the neutron star.

3.3.3 The Uniship Slingshot Trajectory

The slingshot trajectory of a uniship is approximately a hyperbola. As it approaches a neutron star with a distance of closest approach of 100,000 km a uniship it will take approximately 1.5 seconds to circle around the neutron star from an approach distance of 100,000 km to a receding distance of 100,000 km. This time interval is undoubtedly a significant overestimate. Consequently the uniship will only spend a minimal time near the neutron star with little consequent gravitational tidal stresses, magnetic field exposure, and radiation exposure.

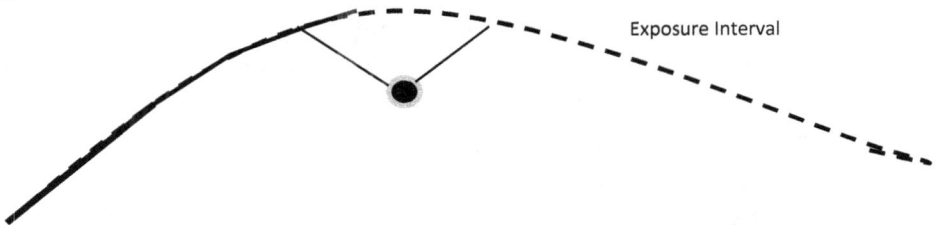

Exposure Interval

Figure 3.2. Depiction of interval of closest approach to the neutron star. The time interval of the uniship in this region is about one second or so. The line segments from the neutron star represent 100,000 km. The uniship speed at the point of closest approach is about 0.7c. When the uniship exits our universe it will disappear from view since electromagnetic radiation (light) from the uniship (or any object) cannot penetrate the boundary of our universe. The dashed line indicates the exit of the uniship from our universe into the Flatverse with Flatverse velocity and momentum components.

Section 5.8 describes the baryon force dynamics of the slingshot mechanism. Having disposed of the problems of a neutron star slingshot we now have the issue of a uniship's escape from our universe. The force of gravity being "confined" to our universe can only bend a uniship's trajectory in our universe. The baryonic force generated by the interaction of a uniship's baryons with the baryons within the neutron universe can cause the uniship's course to be deflected into the Flatverse. It will acquire a momentum in the other twelve spatial dimensions of the Flatverse. Then through a propulsion, and navigation system, that we will describe in chapter 6 and 7 a uniship can proceed to other universes.

4 Alternate Approaches to Escaping our Universe

There are two other possible approaches to escaping from our universe into the Flatverse.[31] One approach sends a uniship through a massive ring or cylinder and uses the baryonic force to enter the Flatverse. The other approach uses a rotating uniship, which of course is primarily made of baryons, to rapidly spin relative to a baryon concentration in its neighborhood thus generating a baryonic force that enables the uniship to enter Flatverse.

Both approaches have some merit. But they also have the drawback of requiring large nearby baryon masses of the order of the sun's mass as well as other drawbacks that make them less desirable than the neutron star slingshot approach described in the previous chapter. Again we defer detailed discussion of the baryonic force to chapter 5.

4.1 Rotating Baryon Ring/Cylinder Mechanism for Escape from Universe

The weakness of the baryonic force requires a large mass, or a gateway ring or a cylinder mechanism to escape from the universe. Alternate mechanisms probably should be located near a large gravitational field – perhaps revolving around a black hole. (See chapter 5 for an explanation.)

We will now consider a rotating ring or cylinder mechanism consisting of a rotating ring or, more likely, a rotating cylinder of great size and length. Not knowing the strength[32] (coupling constant) of the baryonic force we can only be certain that a baryonic ring or cylinder must be of great size, and large baryonic mass. (Most solid substances are overwhelmingly composed of baryons by mass.) Making a ring, or cylinder, rotate at high rotational velocity will require an extremely large amount of energy. But once set in motion the ring or cylinder

---

[31] The Flatverse is the 16-dimensional flat, complex space that contains island universes. Every universe is a surface surrounded by the Flatverse (modulo wormholes connecting universes). Thus we use the term island universes since they exist in the Flatverse "sea." The Flatverse and its island universes constitute the *multiverse* in our terminology.

[32] We estimate the coupling constant in section 5.2. It is extremely small relative to the coupling constants of other forces.

will continue to rotate indefinitely. The rotating ring/cylinder must be electrically neutral. The rotating baryon ring/cylinder does have a baryon charge which generates a baryonic field just as a rotating charged ring generates an electromagnetic field.

The baryon numbers required for a baryon ring or cylinder would have to be extremely large (thus large masses) to generate the required sizeable baryonic "magnetic" field. Thus these mechanisms for uniship entry into the Flatverse are at least many thousands of years into the future.

4.1.1 Spinning Baryon Ring Gateway to the Flatverse

A spinning baryon ring generates a baryonic field that causes a moving baryon uniship in the "center" to spiral into the Flatverse. As soon as the uniship is in the Flatverse, outside our universe, it is no longer visible. Once in the Flatverse the uniship can use its own built-in propulsion system to travel to other universes.

One problem with this approach is the return to our universe. There is no spinning ring in other universes and the construction of one would be costly time-wise, and technology-wise, not to mention the large amount of mass that would have to be assembled to make a ring or cylinder. An alternative approach would be to use the slingshot mechanism for the return to our universe or the spinning uniship mechanism of section 4.2 below.

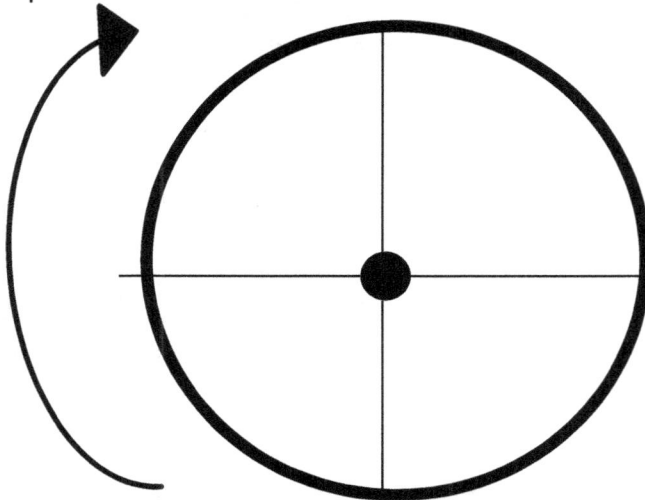

Figure 4.1. A clockwise rotating thick baryonic ring with a uniship at its center moving "straight out of the page."

4.1.2 Spinning Baryon Cylinder Gateway to the Flatverse

A uniship enters a baryonic cylinder and under the baryonic force (See chapter 5.) begins to spiral from our universe into the Flatverse disappearing from view. It then can use its engines to traverse the Flatverse to other universes. It can also view other universes in the Flatverse shining with "baryonic light." The return to our universe would require a baryonic cylinder (or ring or spinning uniship (section 4.2 below) or slingshot maneuver) in the Flatverse to reenter our universe.

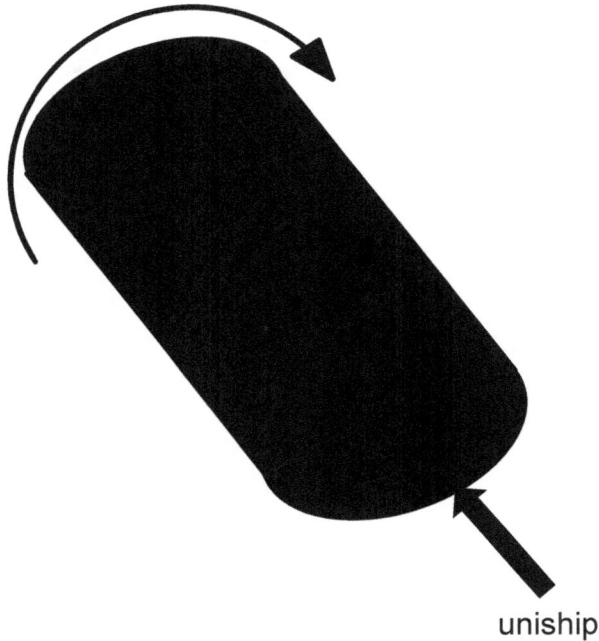

uniship

Figure 4.2. Depiction of a uniship entering a spinning baryonic cylinder along the cylinder's central axis.

Sending a uniship, with a large positive baryonic charge along the axis (center) of the cylinder will cause the uniship to spiral into the Flatverse. The uniship can then travel to other universes.

4.1.3 Rotating Baryon Mass Configuration Gateway to the Flatverse

One can also envision other baryon mass configurations that can be used to enable a uniship to escape from our universe into the Flatverse. One example of a mass configuration is four baryon spheres of large baryon number rotating uniformly. A uniship enters at high speed along their central axis and then spirals into the Flatverse by means of the baryonic force.

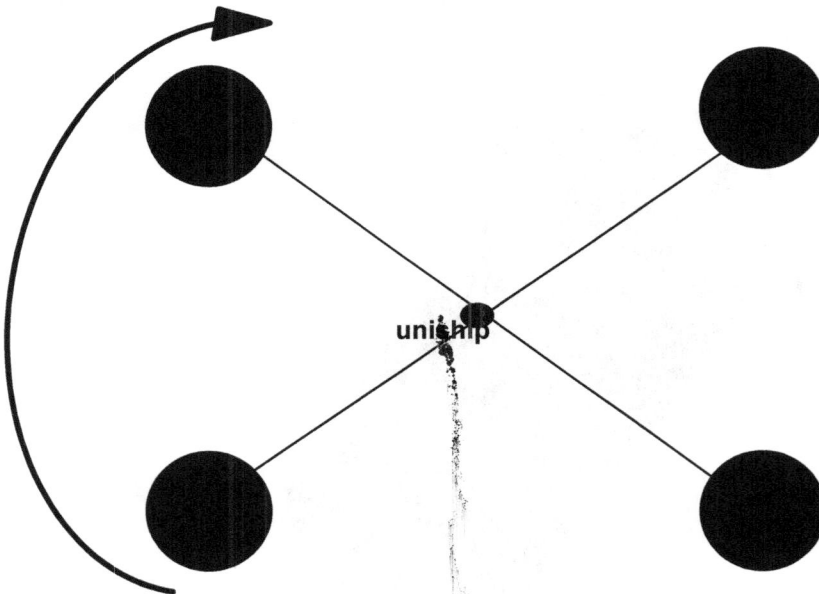

Figure 4.3. Depiction of a configuration of four large masses rotating around a common center. A uniship goes through the center ("out of the page") and enters the Flatverse.

This type of large masses configuration would be difficult to assemble and rotate in synchronization. However it might require less total mass than the ring or cylinder gateways.

4.1.4 Disadvantages of Ring/Cylinder/Mass Configurations to Escape our Universe

The basic problem with all the possible gateways of section 4.1 is that they offer one way transit out of our universe and do not provide for a return.

The approaches in chapter 3 and section 4.2 below provide a capability for a return to our universe.

All of these mechanisms require extremely large masses and are beyond our capabilities in the foreseeable future.

4.2 Escape using a Rapidly Rotating Uniship

In section 4.1 we explored using external large masses to cause a uniship to enter the Flatverse. In this section we consider a rotating uniship moving at high speed through a void in a massive cloud of baryons. It is necessary to move through a void to avoid the severe erosion that a high speed uniship would experience moving through a dense cloud of matter.

Figure 4.4. A uniship traveling outward from page within a void in the midst of a massive baryon cloud.

As the fast rotating uniship moves through a void within the massive cloud of baryons it will experience a baryonic force that will cause it to spiral out of the universe into the surrounding Flatverse. A rotating uniship in a relatively static baryon cloud is equivalent to a non-rotating uniship within a rotating ring or cylinder or mass configuration.

This uniship escape mechanism could presumably be used in any large universe and could also be used to reenter our universe. It avoids the need to create gigantic mass configurations relying on Nature to provide massive clouds within a universe.

The major problem of the rotating uniship is the centripetal force that the crew would experience.

5　The Baryonic "Electric" and "Magnetic" Fields

In Blaha (2014) we described baryonic gauge quantum field theory in detail. In this chapter we will develop the classical theory of the baryonic gauge field. It is similar to electromagnetism with the major exception that it uses 16-dimensional Flatverse coordinates.

5.1 Bary-Electric Fields and Bary-Magnetic Fields

As in electromagnetism there is an antisymmetric tensor of the second rank that appears in the free part of the baryonic field $F_{Bu\mu\nu}(y)$ lagrangian:[33]

$$\mathscr{L}_{Bu} = -\tfrac{1}{4}\, F_{Bu}{}^{ij}(y) F_{Buij}(y) \tag{5.1}$$

where

$$F_{Buij}(y) = \partial B_{ui}(y)/\partial y^j - \partial B_{uj}(y)/\partial y^i \tag{5.2}$$

and i, j = 1, 2, … , 16. The 16th coordinate corresponds to the time coordinate. While the coordinates are complex in general we will treat the 15 spatial coordinates as real and the 16th coordinate as pure imaginary with the resulting invariant interval

$$ds^2 = dy_1{}^2 + dy_2{}^2 + \dots + dy_{15}{}^2 - c^2 dy_{16}{}^2 \tag{5.3}$$

which is invariant under 16 dimensional Lorentz transformations. The coordinates can be transformed into complex-valued coordinates using the Reality group defined in Blaha (2014) and earlier books.

The tensor F_{Buij} is conveniently separated into an baryon electric part and a baryon magnetic part in a manner similar to the separation of the electromagnetic fields into electric and magnetic fields. However the 15 spatial

---

[33] Parts of the following appear in Blaha (2014). They are somewhat modified since we are dealing with the classical, low energy, large distance baryonic field where the quantum coordinate fields Y(y) are well approximated by the classical (non-quantum) Flatverse coordinates y.

dimensions changes the forms of the baryon fields. Analogously, to electromagnetism the baryon force is given by

$$f_i = F_{Buij}(y)J_B{}^j/c \tag{5.4}$$

where $J_B{}^j$ is the jth baryonic current.

The baryon "electric" field is

$$E_{Bui} = -F_{Bui0}(y)/c \tag{5.5}$$

while the baryon "magnetic" field is

$$B_{Bui} = \varepsilon_{ijk}F_{Bu}{}^{jk}(y) \tag{5.6}$$

where i, j, k = 1, 2, ... , 15 and where ε_{ijk} is a totally anti-symmetric tensor with component values ±1. If i < j < k then ε_{ijk} is +1. Even permutations of these three indices yield a value of +1 for the tensor components. Odd permutations of these three indices yield a value of −1. For example, ε_{246} = +1, ε_{426} = −1, ε_{642} = −1, ε_{264} = −1, ε_{462} = +1, ε_{624} = +1.

With these definitions of the $\mathbf{E_{Bu}}$ and $\mathbf{B_{Bu}}$ fields we can easily derive the 16-dimensional generalization of the *Lorentz force law* for a baryon of charge q and 15-velocity v_j:

$$F_i = qE_{Bui} + q\varepsilon_{ijk}v_jB_{Buk}/c \tag{5.7}$$

for i = 1, 2, ... , 15. One important difference from the 4-dimensional case is the forms of the $\mathbf{E_{Bu}}$ and $\mathbf{B_{Bu}}$ fields

$$E_{Bui} = -F_{Bui0}(y)/c = [-\partial B_{u0}(y)/\partial y^i - \partial B_{ui}(y)/\partial y^0] \tag{5.8}$$

or, expressed as a 15-vector,

$$\mathbf{E_{Bu}} = [-\nabla_{15}\phi(y) -\dot{\mathbf{B}}_u(y)]/c \tag{5.9}$$

where ϕ is the baryonic Coulomb potential $B_{u16}(y)$, ∇_{15} is the 15-dimensional grad operator, and $\mathbf{B}_u(y)$ is the baryonic 15-vector potential with the "dot" above it signifying a time (y_{16}) derivative.

The 15-dimensional baryon magnetic field has the form of eqn. 5.6. A specific illustrative case shows the baryon magnetic field exhibits more complexity than the 3-dimensional magnetic field of electromagnetism:

$$B_{Bu1} = \varepsilon_{1jk}F_{Bu}{}^{jk}(y)/c = [F_{Bu}{}^{23}(y) + F_{Bu}{}^{24}(y) + ... + F_{Bu}{}^{215}(y) + F_{Bu}{}^{34}(y) + F_{Bu}{}^{35}(y) +$$
$$... + F_{Bu}{}^{315}(y) + F_{Bu}{}^{45}(y) + ... + F_{Bu}{}^{14,15}(y)]/c \qquad (5.10)$$

Thus each component of the baryon magnetic field impacts on all fifteen spatial directions of the multiverse. For this reason we use spinning rings, mass configurations and uniships to generate baryon magnetic field interactions to enable uniships to escape from our universe's three spatial dimensions. We consider this possibility in more detail in the following sections.

5.2 The Baryonic "Coulombic" Gauge Field

The baryonic gauge field has a "Coulombic" potential part $\phi(y)$, just as the electromagnetic field does. Consequently the total potential between two electromagnetically neutral masses of mass M_1 and M_2, and baryon numbers N_1 and N_2 is

$$V_{tot} = -GM_1M_2/r + (\beta^2/4\pi)\, N_1N_2/r \qquad (5.11)$$

where G is the gravitational constant, and β is analogous to the electric charge e in the electromagnetic Coulomb potential. If both masses are composed of the same substance and have the same mass, then we can set $M_1 = M_2 = M = Nm$ where m is the average mass of the baryons in the masses.[34] In addition we can set $N_1 = N_2 = N$. Then eq. 5.11 becomes

$$V = [-Gm^2 + (\beta^2/4\pi)]N^2/r \qquad (5.12)$$

Note that the gravitational potential term is attractive, and the baryonic potential term is repulsive between baryons.

---

[34] We neglect lepton masses since they are negligible relative to the baryon masses.

In considering eqns. 5.11 and 5.2 we have approximated the baryonic potential with only our universe's spatial coordinates. In reality we should be using the spatial separation in all Flatverse coordinates. However since our universe is close to flat, the distance between two objects that are not too far apart is approximately the same in both coordinate systems. The baryonic potential in Flatverse coordinates is actually

$$\phi(y_1, y_2, \ldots, y_{15}) = (\beta^2/4\pi)N_1N_2/(y_1^2 + y_2^2 + \ldots + y_{15}^2)^{\frac{1}{2}} \qquad (5.13)$$

We now will make an order of magnitude estimate of the baryonic fine structure constant $\beta^2/4\pi$.

5.3 Estimate of the Baryonic Coupling Constant

The baryonic force, and coupling constant, is known to be very small in comparison to gravity and the other known forces. Measurements of the gravitational constant G are significantly different.[35,36] The reason(s) for these discrepancies is not known. We will assume that both the 2010 and 2013 measurements of G are experimentally correct but disagree because of the baryonic force term in eqn. 5.12 that would create a difference in effective G values if the experiments used different masses and thus baryon numbers. Quinn et al found a value for the gravitational constant of $G_1 = 6.67545 \times 10^{-11}$ $m^3kg^{-1}s^{-2}$. The combined 2010 CODATA value for the gravitational constant was $G_2 = 6.67384 \times 10^{-11}$ $m^3kg^{-1}s^{-2}$. Both values are subject to estimated uncertainties.

Suppose these values are correct and due to a difference in the chemical composition (metals) of the test masses used in the experiment. Quinn et all use 1.2 kg test masses composed of Cu-0.7% Te free machining alloy. The CODATA value being a composite of many experiments does not have an effective equivalent test mass value or composition specified.[37] Suppose the test mass value is $N_1^2m_1^2 + N_{1e}^2m_e^2$ for the G_1 result giving

$$-(N_1^2m_1^2 + N_{1e}^2m_e^2)G_1 = [-G(m_1^2N_1^2 + N_{1e}^2m_e^2) + (\beta^2/4\pi)N_1^2] \qquad (5.14)$$

---

[35] T. Quinn et al, Phys. Rev. Lett. **111**, 101102 (2013).

[36] P. J. Mohr, B.N. Taylor, and D. B. Newell, Rev. Mod. Phys. 84, 1527 (2012).

[37] The Eötvös' experiment used a 0.1 gm test mass of RaBr$_2$. R. v. Eötvös, D. Pekár, E. Fekete, Annalen der Physik (Leipzig) 68, 11, 1922.

where G is the real value of the gravitational constant. The total test mass is $(m_1{}^2 N_1{}^2 + N_{1e}{}^2 m_e{}^2)$ with N_1 baryons of average mass m in each test mass and N_{1e} leptons of average mass m_e.

Suppose further the test mass value is $N_2{}^2 m_2{}^2 + N_{2e}{}^2 m_e{}^2$ for the G_2 result giving

$$-(N_2{}^2 m_2{}^2 + N_{2e}{}^2 m_e{}^2)G_2 = [-G(m_2{}^2 N_2{}^2 + N_{2e}{}^2 m_e{}^2) + (\beta^2/4\pi)N_2{}^2] \qquad (5.15)$$

where G is the real value of the gravitational constant. The total test mass is $(m_2{}^2 N_2{}^2 + N_{2e}{}^2 m_e{}^2)$ with N_2 baryons of average mass m_2 in each test mass and N_{2e} leptons of average mass m_e. Since the test masses are electrically neutral and there are approximately equal numbers of protons and neutrons in a test mass it follows approximately that

$$N_{1e} = \tfrac{1}{2}N_1 \quad \text{and} \quad N_{2e} = \tfrac{1}{2}N_2 \qquad (5.16)$$

Subtracting eqn. 5.14 from eqn. 5.15 after some algebra[38] we find

$$\Delta G = -G_2 + G_1 = (\beta^2/4\pi)/(m_2{}^2 + m_e{}^2/2) - (\beta^2/4\pi)/(m_1{}^2 + m_e{}^2/2)$$
$$\simeq (\beta^2/4\pi)(1/m_2{}^2 - 1/m_1{}^2) \qquad (5.17)$$

The masses m_1 and m_2 can differ. For example, if m_H is mass of the hydrogen atom, then $m^{-1} = 1m_H{}^{-1}$ for hydrogen, for carbon $m^{-1} = 1.00782 m_H{}^{-1}$, for copper $m^{-1} = 1.00895 m_H{}^{-1}$, and for lead $m^{-1} = 1.00794 m_H{}^{-1}$.[39] Thus using the Quinn et al and CODATA results and assuming copper and lead test masses we find the order of magnitude *estimate*:

$$\alpha_B = \beta_B{}^2/4\pi \simeq \Delta G/[(1.00895^2 - 1.00794^2)\, m_H{}^2]$$
$$\simeq \Delta G/G\ G\ m_H{}^2/.002037$$
$$\simeq (0.000241/0.002037)Gm_H{}^2$$
$$\simeq .118\ Gm_H{}^2 \qquad (5.18)$$

---

[38] The reduction of the calculation to algebra reminds the author of Nobelist Hans Bethe's remark that he only felt he understood a physical phenomenon when he could reduce it to algebra. This was quite evident when the author collaborated with Professor Bethe on a study of pion condensation in neutron stars some years ago.

[39] "One Hundred Years of the Eötvös Experiment", I. Bod, E. Fischbach, G. Marx and Maria Náray-Ziegler, August, 1990.

indicating a very weak baryonic force consistent with our general view of the multiverse. The baryon fine structure constant is minute in comparison to the electromagnetic fine structure constant $\alpha \simeq 1/137$.

Due to our assumptions in the calculation of α_B which makes it merely an order of magnitude estimate at best we suggest that an experimental group measure G with differing test masses in the same apparatus to obtain a better value for α_B.

5.4 The Baryonic Force on Baryonic Objects

The baryonic force on a moving baryon mass is given by the baryon Lorentz force for a baryon of baryon charge q and 15-velocity v_j:

$$F_i = qE_{Bui} + q\varepsilon_{ijk}v_jB_{Buk}/c \qquad (5.7)$$

for i = 1, 2, ... , 15. The baryonic Coulombic potential is

$$V = N\phi(y_1, y_2, ... , y_{15}) = (\beta^2/4\pi)N/(y_1^2 + y_2^2 + ... + y_{15}^2)^{\frac{1}{2}} \qquad (5.19)$$

where N is the baryon number of the baryon mass. The baryon Coulombic force is

$$F_i = N\nabla_{15i}\phi(y) \qquad (5.20)$$

where ∇_{15i} is the ith component of the 15-dimensional grad operator ∇_{15}.

We shall consider the baryonic force for the various uniship escape mechanisms (discussed earlier) in the following sections.

5.5 Baryonic Force in the Neutron Star Slingshot Mechanism

The baryonic force experienced by a uniship as it travels around a neutron star is the 15-dimensional Lorentz force (eqn. 5.7). It is proportional to the product of the uniship baryon charge q_u and the neutron star baryon charge q_n:

$$F_{uni} = q_uq_nE_{Bui} + q_uq_n\varepsilon_{ijk}v_jB_{Buk}/c \qquad (5.21)$$

The effect of this force is to cause the uniship to loop around the neutron star in our universe's three spatial dimensions and in additional Flatverse dimensions y_i effectively moving the uniship out of our universe as described in section 3.3. It should be remembered that our universe is a surface in the Flatverse and, at any point, a movement off the surface can take the uniship outside our universe in direct analogy to the escape from a 2-dimensional flatland into three dimensions using the electromagnetic force that we describe in section 3.1.

The magnetic term in eqn. 5.21 being dependent on the "cross-product" of the velocity and baryon magnetic field causes the uniship to spiral into the Flatverse. If the velocity is in the "3-direction" $v_j = v\delta_{ij}$ then the force of the magnetic term in eq. 5.21 can be seen by eqn. 5.10 to generate motion in Flatverse dimensions outside our universe. Once outside our universe the uniship is no longer affected by gravity – the baryonic force begins to dominate. (Gravitation ends at the boundaries of our universe[40] – or any universe for that matter – since the total energy of a universe is zero and thus gravitation has no source.)

Thus the slingshot mechanism works using the baryonic force. In particular the baryonic Coulomb force (eqns. 5.19 and 5.20) has components in the Flarverse directions that will cause the uniship to exit the universe as described in section 3.3.3. See Fig. 3.2 for a depiction of the hyperbolic trajectory of the sling shot mechanism. The dashed line part of the hyperbolic trajectory has both spatial coordinates of our universe and other Flatverse coordinate components.

The uniship can then begin to journey in the Flatverse to other island universes – a journey comparable in spirit to the journeys of the great age of exploration in the 15th and 16th centuries that led to the discovery of the Americas and Australia as well as new routes between Europe and the East.

5.6 Rotating Ring/Cylinder/Mass Configuration/Baryon Cloud Baryonic Force for Escape to Flatverse

A rotating ring, cylinder or mass configuration as discussed earlier (Figs. 4.1, 4.2, 4.3 and 4.4) has a baryon current that generates baryon fields and forces (in a manner analogous to a rotating configuration of electric charges generates electromagnetic fields and forces). This baryon current generates a baryon field that will cause a uniship to spiral out of our universe.

The current in the rotating source creates forces between baryon current loops similar to those of the Biot-Savart law in electromagnetism.

---

[40] Our 4-dimensional universe can thus be viewed as a "thin skin" in a 16-dimensional universe.

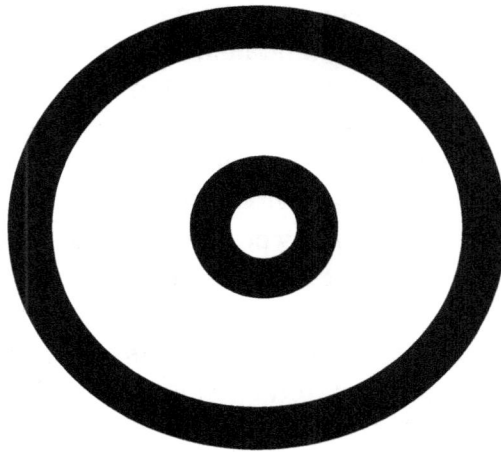

Figure 5.1. A cross section of an outer baryon current loop (in a ring, cylinder, mass configuration, or baryonic cloud) and an inner baryon current loop generated by a large spinning uniship. A "ring within a ring" configuration.

The baryons in the outer "loop" of baryons (ring, cylinder, mass configuration, or baryonic cloud) are effectively fixed in solid masses or slightly moving in a baryon cloud. Thus baryon currents are unlike electromagnetic currents which are generated by moving electrons or ions.

The baryon current generated by the uniship is from the combined rotation of the uniship baryons. It is approximately equivalent to a ring if the uniship is approximately cylindrical.

We have assumed the inner and outer rings are within a region with a large gravitational field such as a region near a black hole or massive star. The presence of a large gravitational field causes the coordinates of the universe x^μ to differ significantly from Flatverse coordinates y^i. Temporarily ignoring quantum aspects, universe coordinates are related to Flatverse coordinates by

$$y_i = f_i(x) \qquad (1.25)$$

Because of the disparity in coordinates the baryonic Biot-Savart force implies a baryonic Ampère force between the currents of the form

$$F = (I_{b1}I_{b2}/c^2) \oint \oint dy_1 \cdot dy_2 \, y/|y|^3 \qquad (5.22)$$

where I_{b1} and I_{b2} are the Flatverse baryonic currents in the rings, $dy_1 \cdot dy_2$ is the spatial 15-dimensional inner product between the line elements of each ring *in Flatverse coordinates*, and **y** is the 15-dimensional spatial vector distance between the line elements.

F clearly has components in Flatverse coordinates that will propel the uniship out of our universe into the Flatverse for transit to other universes. Note the current loops in our universe's coordinates must also be loops in Flatverse coordinates. These loops which might appear circular in a strong gravitational field in our universe will appear to be distorted closed loops in Flatverse coordinates – a simple topological consequence.

Both the uniship and the ring, or other mass configuration, will experience **F** in opposite directions. The uniship, being of much less mass than the enclosing ring, will experience a large acceleration into the Flatverse while the enclosing ring will experience negligible acceleration.

Thus these other mechanisms for escaping our universe are feasible in principle.[41] However the construction of large rings and cylinders in the neighborhood of a black hole or other large mass is a major technological issue that clearly will not happen (if it does) for many millennia. It requires a new massive scale of technology far beyond our present technology.

The neutron star slingshot mechanism appears to be the most promising approach – especially when one realizes that other universes will not have mass configurations similar to those mentioned above except possibly for a large baryon cloud with a central gap located near a black hole or some very massive sun or other massive object.

The major challenge for the neutron star slingshot is the avoidance of difficulties associated with a close approach. This seems possible due to the very short time spent close to the neutron star.

---

[41] The rate of spin of the uniship has to be acceptable from a human factors point of view.

⑥ Multiverse Uniship Designs

In this chapter we will consider design issues for a uniship. Many of the discussion items are also discussed in Blaha (2013a) for much slower speed starships. Within our universe starship speeds of up to hundreds of thousands times the speed of light will enable the wide exploration of the universe. In the case of the multiverse the speeds of uniships should be of the order of hundreds of millions times the speed of light to reach nearby and distant universes. The scale of the multiverse would appear to be trillions of light years. Our goal is to have uniships capable of travel to other universes in (earth) time periods of up to ten years.

6.1 How Many Dimensions Exist Inside an Escaped Uniship?

When a uniship escapes from a universe into the Flatverse the questions arise, What is the dimensionality of the space within the uniship? How do the uniship occupants and equipment interface with the 15 spatial dimensions outside the uniship?

The first issue that must be considered in answering those questions is the physical reality of the Flatverse. Are its coordinates a mathematical fiction? Or are they physically real? The answer seems to lie in the existence of a quantum field that depends directly on the 16 dimensions of the Flatverse. The existence of such a field gives Flatverse coordinates a physical reality since the quantum field, being a physical construct, could not be defined without the physical reality of the Flatverse. An additional requirement for a physical Flatverse is an interaction between the quantum field and the physical mass of our experience. The baryon gauge field $B_{ui}(y)$ that we have postulated previously fulfills both of these requirements. (See Blaha (2014) and earlier parts of this volume.) Thus the Flatverse which is the "ocean" of the set of "island universes" is a physically real entity.

The second issue is the relation of Flatverse coordinates and the coordinates of a universe. We note that every point in a universe's coordinates has a corresponding point in the Flatverse due to the relation:

$$y_i = f_i(x) \qquad\qquad (1.25)$$

Thus every point in a universe has two sets of coordinates. This also holds also for the interior of a uniship.

In the uniship the occupants will live in a 4-dimensional flat space "bubble"[42] within the 15-dimensional Flatverse. The second question we raised concerned how the uniship occupants could interface with 15-dimensional space within and without its confines. This is considered in the next section.

6.2 How Does a Uniship Control Direction and Maneuver in the 16-dimensional Flatverse?

Transiting from a subspace with one set of dimensions to a larger set of dimensions in a larger space is a physically challenging task. One cannot simply use Euclid's algorithm for proving three spatial dimensions (discussed at the beginning of chapter 3). For it assumes that we can move in the direction of any physical dimension. But if the available physical force to move in a given direction is not present then the dimension is physically inaccessible.

In the present case we enable a uniship to escape from our universe using, for example, the slingshot around a neutron star mechanism. After entering the Flatverse it will have some momentum in Flatverse directions beyond the three spatial dimensions of our universe. But that is not enough. A uniship must have the capability to move in any or all fifteen directions in the Flatverse.

Thus we are led to propose a uniship designed to have fifteen thrust exhausts with each direction of thrust pointing in a direction in 15-space. Thrust would be generated in a central chamber and then directed, using (perhaps) bending magnets, through one or more exhaust ports accelerating the uniship in a specified direction. In our universe all fifteen exhaust ports and the direction mechanism would point in a variety of directions in our 3-space. As the uniship slingshots around the neutron star it will experience strong gravitational and baryonic tidal forces. With a correct alignment of the fifteen exhaust ports and central chamber in three dimensions, the exhaust ports and chamber will open

---

[42] An analogous situation is a liquid sphere in zero gravity that splits into two spheres. Both spheres retain the same topology and other characteristics.

like an umbrella in all the directions of the 15 Flatverse dimensions. Then the uniship will be able to accelerate in any direction in the Flatverse.

Fig. 6.1 depicts the uniship before and after the slingshot maneuver.

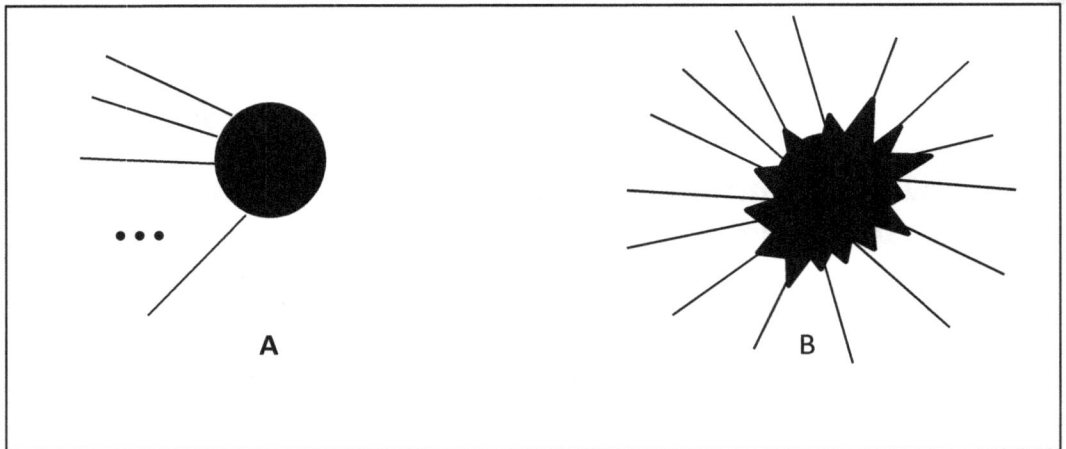

Figure 6.1. Symbolic depiction of the uniship thrust exhaust ports before (A) in 3 dimensions, and in 15 dimensions after (B) the slingshot maneuver to exit our universe.

6.3 Power Source for Uniships

The distance scale of starship travel in our universe (and probably in other universes) between galaxies is of the order of millions of light years. Consequently starships will have to be powered by fusion energy for longer trips and nuclear energy, at minimum, for intra-galaxy trips.[43] Part of the power requirements are necessitated by the need for short travel times of a few years at most so that interstellar commerce, and migration, becomes feasible. For these reasons we wish to have starships capable of speeds up to tens of thousands times the speed of light.

The distance scale of travel in the multiverse is likely to be trillions of light years. We anticipate universes are separated by trillions of light years. Due to these enormous distances, and the fast transportation time we wish, uniships

---

[43] Blaha (2013a) and earlier books.

will require proton-antiproton annihilations on a large scale to power uniships. Proton-antiproton annihilation transforms mass entirely into energy unlike nuclear and fusion reactions where only a fraction of the masses of the particles are converted into energy yielding of the order of 1,000 the energy production of fusion.

Our power requirements are based on a need for fairly short travel times of a few tens of years at most so that inter-universe travel does not exceed a few tens of years. For this reason we wish uniships to be capable of speeds up to millions of times the speed of light.

6.4 Possible Uniship Design

We therefore require a large sphere containing protons that are densely packed. (An alternative that would avoid electromagnetic repulsion problems associated with densely packed protons would be a densely packed sphere of hydrogen atoms.) We also need a large sphere containing antiprotons that are densely packed. (Again an alternative that would avoid electromagnetic repulsion problems associated with densely packed antiprotons would be a densely packed sphere of anti-hydrogen atoms.) In addition to being the most powerful energy source available these spheres would form a baryon dipole that could play a role in the escape of a uniship from a universe.

We therefore start with a uniship design of the form of Fig. 6.2.

Figure 6.2. An initial depiction of the overall design of a uniship. The upper and lower spheres contain a dense mass of protons and of antiprotons respectively. The antiproton sphere requires a confinement mechanism to avoid premature particle annihilation. (The spheres could alternatively contain liquid hydrogen and anti-hydrogen to avoid electromagnetic repulsion issues.) The module in front contains the crew quarters, equipment and cargo. The tail section contains a central region that controls the particle-anti-

particle combustion process directing the thrust through one or more of the fifteen thrust chambers to power the uniship in the desired direction in the Flatverse.

6.5 Intra-Universe Starship Designs Summary

Starship design has a number of requirements to satisfy for trips within the galaxy and a more stringent set of requirements for travel to other galaxies in our universe. Uniships have a yet stricter set of requirements for travel to other universes within the multiverse. The discussion in this section sets the background, based on our starship designs, for the consideration of important design details for uniships presented in succeeding chapters. Most of this section is abstracted from Blaha (2013a) and earlier books on nuclear ships, space guns, and starships. Appendix A briefly overviews the designs in these earlier books for mass transit from the earth to space, travel in the solar system, and, most importantly, starship travel within our galaxy and to other galaxies.

6.5.1 Types of Starships

There are two general types of starships due to the nature of starship drives: starships based on circular accelerators and starships based on linear accelerators. In both cases we assume the process that creates the quark-gluon plasma thrust is a stream of collisions of spherules of some material.

6.5.2 Starship Engine Energy Sources

The starship engine that we have designed requires massive amounts of energy. At this point in time the only feasible energy source for starships is nuclear energy. It is reasonable to expect that fusion energy, a more concentrated energy source, will become a reality within the next thirty years. In either case the starship will need an energy source to drive the spherule accelerator rings and associated devices of the engine for periods up to perhaps a few months or years, then turn off for perhaps many years, and then resume operations for further maneuvers.

In the extreme case of travel to another galaxy, the energy source will need to turn off for up to millions of years of starship time. While the energy source is turned off, a residual "battery" will need to operate to support monitoring the progress of time, activating the startup of the main energy source, and possibly to detect and monitor objects ahead of the starship in the line of flight. This battery source may well be a plutonium (or longer lived) source similar to those used in current space probes.[44]

The main energy source, if it is a nuclear reactor of some kind, will probably have to be a reactor that is different from current nuclear reactors. Chapter 5 of Blaha (2013a) describes long shelf life nuclear reactors that could meet this need. Since the startup process from a battery driven state needs to be gradual due to a "small" battery, it appears the set of nuclear reactors would be composed of perhaps five reactors of increasing size. The battery starts the smallest reactor by concentrating its nuclear fuel. The smallest reactor then generates the energy to concentrate that fuel, and start the second smallest reactor, and so on until the main reactor(s) is started. At this point the accelerators power up and starship thrust begins.

If the source of the energy is fusion energy then the startup process might begin in small stages in the boot up of the fusion reaction through fusing larger and larger amounts of (perhaps) ^{3}He with increasingly powerful laser beams a la the tokamak approach.

During the coasting period of a starship, nuclear reactors should be powered down to conserve nuclear fuel. When powering down a nuclear reactor power source the nuclear material (U^{235} or plutonium) (the reactor fuel) residing in the medium of the reactors would be diluted to sharply reduce fission reactions to "near zero" using the energy of the next largest reactor. The smallest reactor would be powered down by a battery. This battery would retain enough energy to bring the smallest reactor back up after the coasting period ends. Then the reactors would boot up in turn to provide energy to the vehicle. (The battery would be at extremely low temperature during a coasting phase and thus not lose a significant amount of electrical power.)

In the case of a fusion power source a battery could be used to initiate the fusion power. A gradual turnoff process could execute at the start of a coasting phase to bring the fusion process to zero in such a way that the battery could initiate the boot up process for the fusion power source at the end of a coasting period.

---

[44] We note that a natural nuclear reactor existed in the Congo Region of Central Africa for millions of years. (Parenthetical note: Could this be the stimulus for the rapid evolution of species in Africa including very early Mankind?)

6.5.3 An Alternate Starship Accelerator Ring Engine Design

In an earlier work Blaha (2009b) we proposed an alternate accelerator ring design for quark-gluon fluid acceleration. It appears that this design is not feasible with current or near term (one hundred years) technology. The reasons are:

1. The quark-gluon fluid ring is not feasible because fireballs cannot be injected and accelerated to form a ring in a few fm/c - the time available.

2. A quark-gluon fluid ring would be similar to fusion tokamaks but much more challenging in its requirements for confinement and stability due to the much higher density and temperature of a quark-gluon fluid ring.

6.5.4 Some Starship Configurations

There are a variety of possible configurations for faster than light starships. In this subsection we categorize the starship configurations by their thrusters.

6.5.4.1 Rear Thrust Exhausts on Circular Accelerator Starships

In Blaha (2010a), and later books, quarks and gluons were shown to have complex 3-momenta. The real part of the 3-momenta of each quark was orthogonal to the imaginary part of its 3-momenta.

If we "add" pairs of exiting quarks to create a complex 3-momentum to the rear, then a complex total pair 3-momentum is created that generates a rearward thrust. The complex thrust can be used in a one thrust exhausts, or two thrust (or multi-thrust) exhausts as shown in Figs. 6.3 and 6.4.

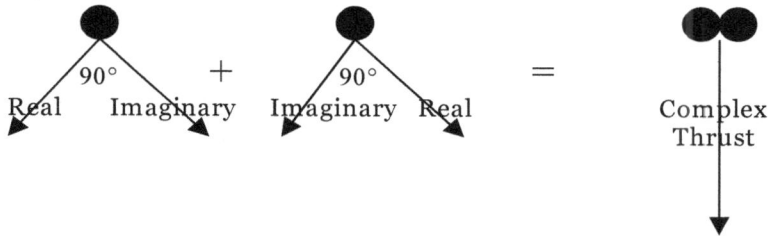

Figure 6.3. A circular accelerator starship with one thrust exhaust.

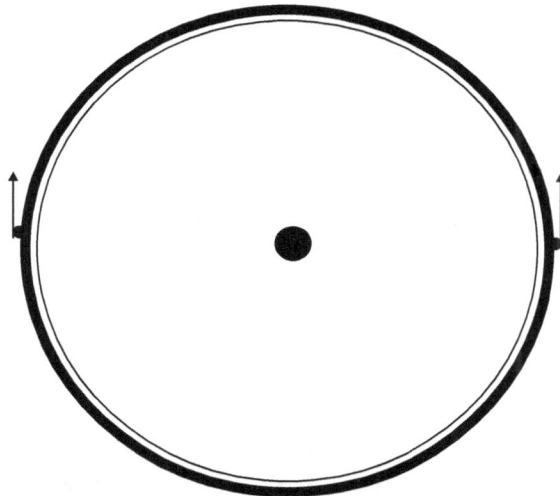

Figure 6.4. Top view of a circular starship with two complex thrust exhausts.

These types of starships generate thrust that only moves the starship forward in space. They do not cause the starship to rotate. The next subsection introduces the use of the imaginary perpendicular part of a starship's thrust to cause the starship to rotate.

6.5.4.2 Rotating Circular and Cylindrical Starships due to Imaginary Part of Thrust

In the case of disc-shaped and cylindrical starships the imaginary part of the quark-gluon thrust can be directed by magnets to be tangent to the circular edge of the starship. The resulting tangential force causes the starship to rotate through an imaginary angle. The rotation creates a *negative* centripetal force in the starship – "artificial gravity" that varies with the distance from the central axis of the starship. Fig. 6.5 shows a "top view" of a starship.

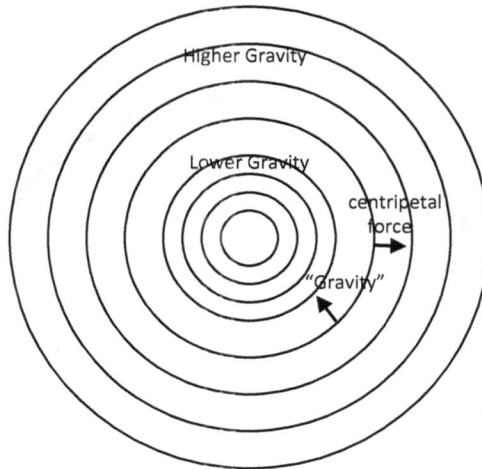

Figure 6.5. Top View of a Disc-like or Cylindrical Starship. View of cargo/people levels of inner hub. The center has lower "gravity." Circles indicate levels of equal artificial gravity in a disc-shaped or cylindrical starship. The outer parts have higher "gravity." The "gravity" force is inward towards the center on all levels. The centripetal force is outward from the center as shown by the arrow in the diagram and the discussion below (opposite to the direction of conventional centripetal force.) The artificial "gravity" force is thus inward to the center.

Fig. 6.6 is an example of a disc-shaped starship with the thrust direction displayed. The horizontal thrust arrows represent the real part of the quark-gluon thrust. It accelerates the starship to the left in the figure. The vertical arrows represent the imaginary part of the thrust. It rotates the starship in a counterclockwise direction.

Figure 6.6. A disc-like starship with thrust consisting of a real and imaginary part. The real part drives the ship to the left. The imaginary part is tangent to the edge of the starship causing it to rotate counterclockwise.

6.5.4.3 Artificial Gravity Levels - Rotating Circular and Cylindrical Starships

In rotating starships the real part of the thrust propels a starship. The combination of the real and imaginary parts of the thrust enables a starship to exceed the speed of light. The imaginary part of the thrust causes the starship to spin around its central axis.

We will examine the case of a cylindrical starship and see how the "artificial gravity" emerges as a result of the imaginary part of the thrust. Fig. 6.7 shows a cylindrical starship and Fig. 6.5 shows the cylindrical coordinate system used to calculate its motion.

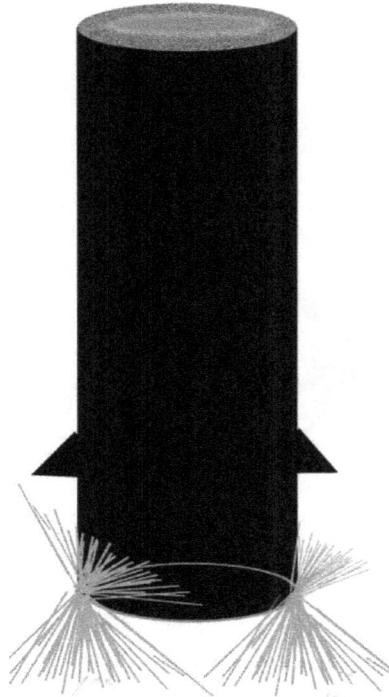

Figure 6.7. Cigar shaped starship with "horizontal accelerator ring(s). The real part of the thrust points downward. The imaginary part of the thrust is horizontal and tangent to the cigar surface. The fins are for supplementary nuclear maneuvering rockets.

The force generated by the quark-gluon thrust has the form

$$\mathbf{F} = g_r \check{\mathbf{z}} + i g_i \boldsymbol{\phi} \tag{6.1}$$

in the starship's cylindrical coordinates rest frame where $\check{z}$ is a unit vector in the positive z direction and $\boldsymbol{\phi}$ is a unit vector in the positive ϕ angle direction. g_r and

g_i are constants specifying the thrust. g_r is the starship's mass times its real-valued acceleration in the ž direction. g_i is the starship's mass times its imaginary-valued acceleration in the ø direction. The time derivative of the momentum **p** of a small part of mass m_0 at the edge of the starship is

$$d\mathbf{p}/dt = m_0(\gamma z')'\mathbf{\check{z}} + m_0[(\gamma\rho\emptyset')' + \gamma\rho'\emptyset')]\mathbf{\emptyset} + m_0[-\gamma\rho(\emptyset')^2 + (\gamma\rho')']\mathbf{\rho} \qquad (6.2)$$

where **ρ** is the unit vector in the positive ρ direction and the ′ symbol indicates a time derivative.

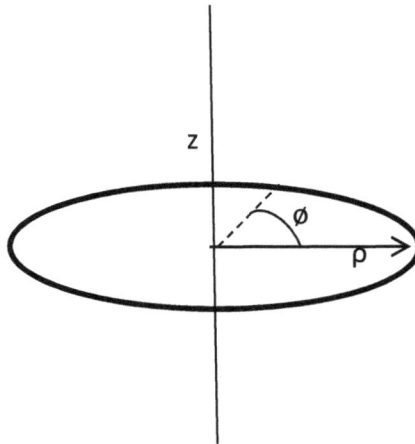

Figure 6.8. Starship cylindrical coordinate system. The cigar shaped Starship in Fig. 6.9 is centered on the z axis.

If we set $\rho' = 0$ since there is no force component in the **ρ** direction and the starship is not contracting, then setting

$$\mathbf{F} = d\mathbf{p}/dt \qquad (6.3)$$

implies

$$m_0\rho(\gamma\emptyset')' = ig_i \qquad (6.4)$$

The centripetal force is

$$F_{centripetal} = m_0\gamma v_c^2/\rho \qquad (6.5)$$

where v_c is the rotational velocity of the mass. It satisfies

$$v_c = \rho\phi'$$ (6.6)

Since ϕ' is imaginary by eq. 6.4, v_c Is imaginary and thus the centripetal force $F_{centripetal}$ is negative by eq. 6.5. The centripetal force is "opposite" to the centripetal force normally experienced – away from the center. The artificial "gravity" force – really the opposite of the centripetal force – is towards the center. See Fig. 6.7.

The artificial "gravity" engendered by the rotation raises an issue. The acceleration of the rotation implied by eqns. 6.2 and 6.3, if continued, would generate an enormous inward "gravitational" force crushing the starship's occupants. This problem can be solved by causing an oscillation in the rotation between clockwise and counter-clockwise directions by repeatedly flipping the value of the imaginary force g_i (eq. 6.1) to maintain a constant rotation speed (and thus have constant values of "gravity.")

A circular (disc-shaped) starship can also rotate to create artificial gravity. Fig. 6.6 illustrates a rotating disc starship, Again the gravitational force is towards the center with lower gravity in the central region. The disc moves to the left due to the real thrust.

6.5.4.4 Starships Based on Linear Particle Accelerators

Starships based on linear accelerators can have a variety of forms. Fig. 6.9 contains a diamond shaped starship. Generally starships based on linear accelerators need great length – of the order of miles for the primary linear accelerator tubes unless a very rapid linear accelerator mechanism is developed.

The angles between the linear accelerator tubes are determined by maximizing the efficiency, and amount, of thrust generation. Thus the shape of the *lower half* of the diamond-shaped starship is more or less determined. The lower part of the starship can be sharp edged diamond shaped or rounded diamond shaped. The upper part of the starship should minimize the effects of space dust.

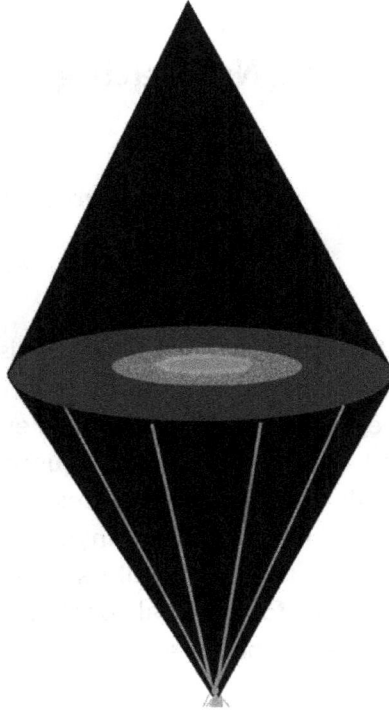

Figure 6.9. A diamond shaped ship powered by four linear accelerators. The lower part has the accelerators, magnets, nuclear reactors, and propellant. The upper part contains the crew, cargo, and shielding as well as nuclear shuttles for exploring a solar system.

7 Traveling and Navigating in the Multiverse

I saw Eternity the other night,
Like a great ring of pure and endless light,
All calm as it was bright;
"The World" – Henry Vaughan

Upon entering the multiverse the first uniship will want to scan the universes and simply create charts of "nearby" interesting universes rather like astronomers on earth began by charting the galaxies in the heavens and looking for novel features and phenomena. Subsequently, the first uniship, and subsequent uniships, will travel to universes of interest and begin their exploration. What they will encounter is anyone's guess. But we have good reasons to believe that other universes will have similar physical laws although they may have different values for physical constants such as elementary particle masses and coupling constants. The reasons are:

1. We believe the dimensionality of universes is fixed by the principles of Asynchronous Logic that make four space-time dimensions the minimal acceptable number of dimensions. (We note that higher dimensional space-times are not excluded by these principles. But Nature tends to favor extremums – a minimal number of dimensions in the present case. And Nature also tends to repeat successful designs.)

2. We have shown that complex four dimensional space-time leads to The Standard Model directly (particularly the form of the fermion mass spectrum) with an additional $SU(2){\otimes}U(1)$ that we associate with the Dark Matter sector.

3. Since all space and time measurements yield real-valued numbers we require a Reality group to map complex space-time coordinates to real-valued coordinates. The Reality group gives the group structure of an extended Standard Model $SU(3){\otimes}SU(2){\otimes}U(1){\otimes}SU(2){\otimes}U(1)$. Thus the form of the Standard Model (somewhat extended) is directly based on Logic and geometry.

4. Gravitation is also based on geometry.

Thus we have the form of physical theory based on geometry although the numerical constants in the theory remain to be specified. We conclude almost every universe has a similar fundamental physics theory possibly with different physical constants. We say "almost every" because the possibility of irregular universes cannot be ruled out. Nature can choose the bizarre at times. In Blaha (2014) we describe the types of universes that can occur. In earlier books such as those listed in the Preface we developed the geometric theory of Physics described briefly above.

We have described the necessary structure of universes realizing that they may contain unusual phenomena and may contain alien civilizations with which we can have scientific and cultural exchanges. Consequently it behooves us to explore the multiverse to extend the range of human knowledge and possibly find reasons for establishing commerce.

In this chapter we will consider aspects of uniship travel, exploration, and navigation. In the following chapters we will consider specific features required of a uniship.

7.1 Uniship Range and Travel Times

Ideally we would like uniships capable of traveling large distances of the order of trillions of light years in relatively short times of the order of years or, at least, decades. These wishes cannot be accomplished with current or foreseeable technology. Crews can only withstand modest accelerations up to 8g at present and probably not much more with foreseeable space medicine advances. Thus accelerating to millions of times the speed of light will take a considerable amount of time.

Given the acceleration time requirements trips between universes become lengthy – perhaps decades of earth years. On a uniship, at these very high speeds the occupants and equipment will see time progress rapidly – at a rate equal to the speed measured in units of c (the speed of light) times earth time because of relativitistic dilation. A uniship traveling at 1,000,000c will experience time increasing at 1,000,000 times earth time. An earth year thus becomes a million years of uniship time. This time dilation effect places stringent requirements on uniship equipment and requires suspended animation for the crew. We will discuss these issues in more detail later.

On the positive side, if the uniship makes a roundtrip at that speed only two earth years will have elapsed, and if they were in perfect suspended animation, the crew's body clocks will have aged only two years in sync with earth time.

The preceding discussion has ignored one major point – the energy required to accelerate to millions of c is enormous. That is why we have to use the most powerful "concentrated" energy available. We hope that energy derived in quantity from particle annihilation will be available 50,000 years hence if not much sooner. Otherwise travel into the multiverse becomes analogous to an ant swimming across the Pacific Ocean.

The above comments cannot be viewed as encouraging. But with scientific and technological advances in the next 50,000 years there is hope. And there is good reason to begin thinking of that ultimate goal and how we may eventually reach it.

7.2 Uniship Baryonic 16-Dimensional Observation/Seeing Techniques

If a uniship enters the Flatverse of the multiverse there should be countless universes in view if there is a method for seeing them. However the ability to see/detect universes is very limited. We cannot detect gravitational waves from universes because the universes have total energy zero and thus there is no source for gravity waves to be generated.

More importantly we cannot detect electromagnetic waves because they are confined to the universe within which they were created and, at best, simply circulate through that universe endlessly.

Consequently, the only radiation that we can expect to encounter in the Flatverse is baryonic radiation from universes. There are no other known long range types of radiation. This creates a quandary that we can only begin to address at our present state of knowledge. Observing baryonic radiation will be a difficult task – not just because of the weakness of the baryonic field coupling constant but also for several important reasons:

1. We can't see baryonic radiation either visually or through technology at present. No detectors.
2. We cannot focus on the source(s) of baryonic radiation so as to distinguish their direction and distribution.

3. We have no mechanism to magnify the pattern of incoming baryonic radiation. No lenses, telescopes or other viewing mechanisms with "zoom" capabilities. And no technology to amplify baryonic radiation.

4. Baryonic radiation is 16-dimensional. We would have to be able to create 3-dimensional hologram projections that provide an intelligible view. The holographic images would have to be manipulated to enable navigation.

5. Baryonic radiation has 14 polarizations which would provide information on the nature of universes. The detection and analysis of these polarizations is currently beyond our capabilities.

Our inability to detect gravity waves (except in recent studies of the Big Bang – BICEP2) shows the difficulty of detecting and analyzing baryonic radiation.

7.2.1 Relativistic Effects on Baryonic Radiation

The baryonic view of the multiverse that a uniship crew "sees", when the uniship is traveling faster than the speed of light, is very different from its view when traveling at low speeds of a few tens of miles per second.

An observer on a uniship traveling at a relativistic speed near, but below,

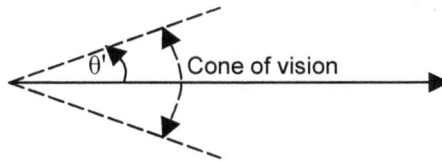

the speed of light will detect universes with baryonic radiation compressed to within a cone in the frontal direction of the uniship (Fig. 7.1). The baryonic radiation cone becomes narrower as the speed of light is approached due to aberration and in the limit as the speed approaches the speed of light becomes a point directly ahead of the uniship.

Figure 7.1. Baryonic radiation cone of visibility around direction of uniship motion in the uniship coordinate system with the angle θ' determined by eq. 7.1 for sublight uniship speeds.

The relativistic equation for baryonic radiation aberration is

$$\cos \theta' = (\cos \theta + \beta)/(1 + \beta \cos \theta) \tag{7.1}$$

where θ is the angle of a universe relative to the uniship's direction of motion as measured in the Flatverse coordinate system and θ' is the angle of a universe relative to the uniship's direction of motion as measured in the uniship's coordinate system.

The inverse relation is

$$\cos\theta = (\cos\theta' - \beta)/(1 - \beta\cos\theta') \qquad (7.2)$$

7.2.1.1 Sublight Case: β < 1

As β → 1 (the speed of light) eq. 7.1 indicates θ' → 0° showing the entire view of the multiuniverse baryonic radiation is compressed to the forward direction. Fig. 7.1 shows the cone of visibility for a uniship traveling near the speed of light at perhaps .6c - .9c. The cone angle θ' satisfies

$$\cos\theta' > \beta \qquad (7.3)$$

The rest of the field of view of the uniship is total baryonic radiation blackness except the point in the directly rearward direction (θ' = 180°) for any object at θ = 180°.

7.2.1.2 Superluminal Case: β > 1

For β > 1 eqns. 7.1 and 7.2 still hold and there is a cone of baryonic radiation visibility similar to that depicted in Fig. 7.1. However the cone angle θ' for superluminal speeds, β > 1, satisfies the relation

$$\cos\theta' > 1/\beta \qquad (7.4)$$

The rest of the field of view of the uniship is total baryonic radiation blackness, as in the sub-light speed case, except the point in the directly rearward direction (θ' = 180°). We note that as β gets very large the cone of visibility becomes larger. At β = ∞ the cone of visibility becomes the angular region between θ' = 0° and θ' = 90° (the forward hemisphere).

7.2.1.3 Superluminal Uniship Visibility

As a result, "visual" baryonic radiation navigation at high superluminal speeds becomes difficult unless we develop an electronic imaging system that "undoes" the effects of aberration and enables "visual" baryonic radiation navigation.

A further problem is the location of a destination. If we send a uniship to a far universe we have to project the location of the universe at the time the uniship arrives based on the universe's current motion. If the motion of the universe is modified by baryonic forces exerted by nearby universes as baryonic radiation from the destination universe was traveling to the uniship, or if the universe's motion is not accurately determined, a uniship could arrive at a point that is still some distance from the destination universe. Thus navigation to a far universe is a significant issue.

7.2.1.4 Effect of Doppler Shift at Superluminal Speeds

A uniship traveling at relativistic sublight speeds will see universes having their baryonic radiation spectrum (color) changed significantly due to the Doppler Shift effect. At superluminal speeds the Doppler Shift will also change the "colors" of objects "seen" by the uniship.

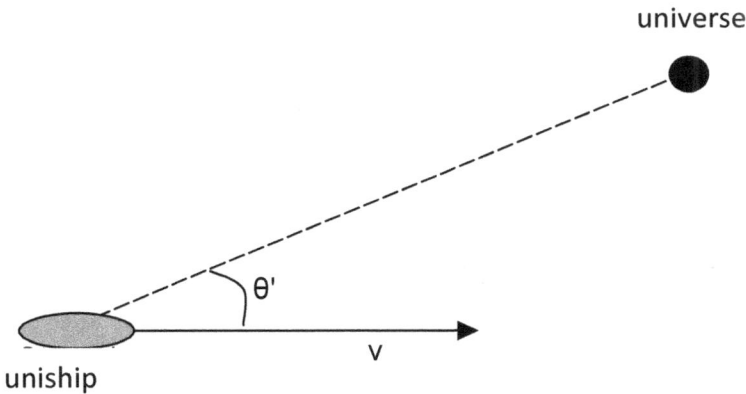

Figure 7.2. The angle of a universe θ' with respect to the uniship's velocity v.

This issue is again surmountable if we use electronic imaging techniques to "undo" the Doppler shift and thus display universes with their true baryonic radiation spectrum in the uniship's reference frame.

The relativistic Doppler shift for sublight speeds of a baryonic radiation wave of frequency v is given by

$$v = v_0(1 - \beta^2)^{\frac{1}{2}}/(1 - \beta \cos \theta') \tag{7.5}$$

where v_0 is the frequency of the baryonic radiation emitted by the source universe and θ' is the angle of the source relative to the uniship's velocity (Fig. 7.2). The Doppler shift for superluminal speeds is

$$v = v_0(\beta^2 - 1)^{\frac{1}{2}}/(\beta \cos \theta' - 1) \tag{7.6}$$

This can be seen by considering a baryonic plane wave, which is a combination of

$$\cos[(k \cdot x - vt)/2\pi] \quad \text{and} \quad \sin[(k \cdot x - vt)/2\pi] \tag{7.7}$$

Upon transforming from the uniship coordinate system, for example, to a coordinate system moving in the "x-direction" at a speed faster than light, both the energy v' (up to a constant) and the time t' obtain a factor of i (that cancel each other) so eq. 7.6 is the correct frequency in the superluminal (faster than light) frame. The sign of the frequency is always positive by convention due to the form of baryonic waves and eq. 7.4 dictates the form of the denominator in eq. 7.6.

For large $\beta \gg 1$ eq. 7.6 becomes approximately

$$v \approx v_0/\cos \theta' \tag{7.8}$$

In the forward direction $\theta' = 0$ the Doppler shift goes to zero. Due to eq. 7.4 the maximum value of the Doppler shift for large β in the field of vision is

$$v \approx \beta v_0 \tag{7.9}$$

So the "wide" angle baryonic waves are shifted to large frequency.

Eq. 7.6 and the discussion that follows suggests that frequency shifts will be substantial for extremely fast uniships. The result will be a distorted view of the multiverse.

However electronic imaging techniques can be implemented to restore the "correct" view of the baryonic radiation. The combined effects of aberration and the Doppler shift on the view of the multiverse from the uniship can be electronically corrected to give a "normal" view of the multiverse. In addition a projection system, probably based on holograms, is required to transform 16-dimensional views of the multiverse into sets of 3-dimensional depictions of parts of the 16-dimensional view.

7.3 Uniship Navigation

Navigating on earth and in space is often a difficult task. First one must know where one is and then one must know where the destination is, and how to get there. In the multiverse all three items are challenging to discern. For we are in a 16-dimensional space where our intuition, based as it is on three dimensional space, fails. We are thus at the mercy of technology to detect these three things with only electronic baryonic eyes to see and guide our motion.

Baryon detectors on board detect other universes through their baryonic radiation just as we detect stars and galaxies by their electromagnetic radiation currently. Within the next 50,000 years we anticipate that baryonic radiation optics will develop and mature to the point where it can provide visual capabilities similar to electromagnetic light that we use at present. Then we can develop universe maps for the multiverse just as we have star maps currently.

An important issue is the ability to distinguish anti-matter universes from universes dominated by matter such as our universe. We do not want uniships to enter anti-universes unless we can properly shield them from disintegrating under particle-antiparticle interactions.

The navigation system, using 3-dimensional views (holograms) obtained from 16-dimensional pictures of the multiverse, can then plot courses to universes of interest for exploration.

The courses selected then direct the 15-directional thrust system to execute the correct combination of accelerations to travel to the selected universe.

Most of the technology that we have discussed remains to be created. However the rapid progress of technology, if it continues, would seem to be able to provide the needed components in the future.

⑧ The Horizons of Universes and Uniships

We have portrayed universes as islands in the Flatverse. A universe is a surface in the 16-dimensional Flatverse with a well-defined border in the Flatverse (up to quantum effects) defined by

$$y_i = f_i(x) \tag{1.25}$$

for i = 1, 2, ... , 16. The 4-dimensional coordinates x form an infinite domain that specifies a surface in the Flatverse. We can view the surface as a type of horizon. Inside the horizon are the contents of the universe and two sets of coordinates, Flatverse coordinates and the gravitationally curved coordinates of the universe. Outside the horizon is flat 16-dimensional space with zero gravitational effects and zero electromagnetic radiation emanated by the universe. (The curvature of a universe confines all electromagnetic radiation to within the universe.) The only remaining long range interaction is the baryonic interaction as we pointed out in previous chapters. This interaction "penetrates" the horizon and can interact with baryons on either side of the horizon.

The horizon of a universe is thus permeable for baryonic radiation. We now have to consider what happens when a uniship leaves or enters a universe.

8.1 Uniship Exit from a Universe

We have seen that only baryonic radiation can penetrate a universe horizon. We must now amend that conclusion to consider the exit or entry of uniships from a universe. In the slingshot mechanism for exiting, and the other possible exit methods that we considered, a key role was played by the baryonic interaction in all cases. The role is easily visualized by considering the example of escaping from 2-dimensional flatland using the magnetic force (section 3.1).

However in the current situation (4-dimensional object exiting into a 16-dimensional space) we have to inquire into the nature and results of the exit. It is clear that the escape of a uniship must be view as a bubble (the 4-dimensional universe) subdividing into two 4-dimensional bubbles in 16-dimensional space in a continuous process rather like biological cell division. One bubble is of course the universe. The other bubble is the uniship which is also a surface from the viewpoint of the Flatverse. Fig. 8.1 symbolically depicts an intermediate stage in the subdivision process.

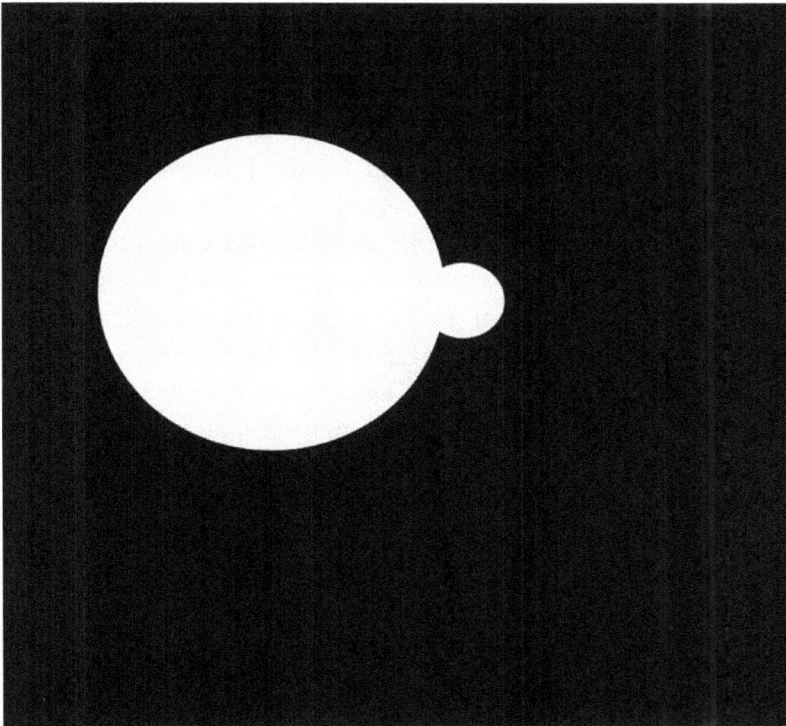

Figure 8.1. Symbolic depiction of an intermediate stage in the subdivision representing the exit of a uniship from a universe. The uniship (small circle) emerges from the universe (large circle) into the Flatverse. (Not drawn to scale.)

We thus see that a uniship is a 4-dimensional entity but with fifteen thrust ports extruding into 15-dimensional space. (The thrust ports were generated by tidal effects in the slingshot process.) Thus it is a hybrid with 4-dimensional and 16-dimensional parts. The crew, the equipment and the fuel modules will be in 3-dimensional space. The combustion chamber and thrust ports and viewing mechanism for navigation will be in 15-dimensional space. The combined parts will constitute the uniship. (We note that the 3-dimensional parts are in a subspace of 15-dimensional space just as a piece of paper is a 2-dimensional object in 3-dimensional space.) Therefore there are no connection problems between the 3-dimensional and 15-dimensional parts of the uniship.

8.2 Uniship Entry into a Universe

We now consider the entry of a uniship into a universe, which we will assume is 4-dimensional like our universe for reasons given earlier (section 1.3). The uniship has both 3-dimensional and 15-dimensional spatial parts. If we position the uniship 3-dimensional part within the universe and retract or rotate the 15-dimensional part into the 3-dimensions of the universe, then the uniship will be entirely within the universe giving a successful entry.

Retracting 15-dimensional parts is easy to visualize but requires a mechanism for retraction in 12 of the 15 dimensions. This mechanism can be a simple retraction mechanism but must exist in each of the dimensions outside the universe being entered. Rotating thrust exhausts also is easy to visualize but, again, the rotation device must be in the 12 external dimensions and rotate into the three spatial dimensions of the universe being entered. In both cases we are faced with Cheshire cat situations: after the retraction or rotation the 12-dimensional devices that perform these chores will still exist and thus not be fully within the destination universe.

Whether the uniship can explore the new universe being partly outside it is an open question. This author feels that the uniship will be able to successfully navigate in the new universe rather like a shark swims the sea with its fin sticking out of the water. The 12-dimensional "fin" will add to the mass of the uniship but will not be affected by gravitation because it is out of the universe.

⑨ Issues for Life on a Multiverse Uniship

Life on a uniship that travels to universes is similar to life on a starship that travels to stars and galaxies except it has significantly more stringent requirements. Travel within the universe is measured in light years ranging to millions of light years. Travel within the multiverse is most likely in hundreds of billions to trillions of light years in starship time. This results in massively large requirements for fuel, materials strength and longevity, and in human suspended animation time among other larger requirements.

Starship requirements are described in Blaha (2013a) and his earlier books. In this chapter we will describe some uniship requirements. The following chapters and appendix B describe other uniship requirements.

9.1. Long Distance Starship Requirements for Travel to Far Stars and Galaxies

If we wish to travel to long distances – up to trillions of light years eventually, then critical advances are necessary that could take up to 50,000 years.

We see the uniship effort as a long term exploration and colonization program in an ever widening ring around our universe. In this chapter we discuss many of the advances that would be needed.

A major problem of uniships is the rapid progress of time on a much faster than light uniship. If a uniship has a speed that is much faster than the speed of light, then the progress of time in the uniship is much faster than the progress of time on earth.[45] For example if the uniship is traveling 5,000,000 times the speed of light, then the increase in time on the uniship is 5,000,000 times the increase in time on earth. In an interval of one year of earth time, 5,000,000 years will have passed on the uniship.

The extraordinarily fast passage of time on a very fast uniship requires materials, equipment and engines to continue to work effectively for long periods of uniship time which is just as real on a uniship as earth time is real on

---

[45] See p. 15 of Blaha (2011c): *All The Universe.*

earth. One cannot avoid the fact that the distance traveled by a uniship measured in light years is equal to the time of flight to cover that distance measured in years.[46]

Thus we come to the first important long distance uniship requirement – very long lifetime equipment and uniship superstructure. Other requirements follow in this chapter.

9.2 Long-Lived Materials

The materials that we use today to build large vehicles such as oil tankers, submarines and aircraft carriers are meant to last up to, at most, a century and often much less. Many of these materials age, deteriorate, rust, migrate within computer chips over time, and actually slowly flow like a liquid in many cases.

Not many materials keep their original characteristics over long periods of time. In the past fifty years there has been much progress in developing new harder, stronger and age resistant materials and metals. But uniships requirements are extraordinarily larger.

Uniships, in which time moves quickly so that thousands and perhaps millions of years of uniship time elapse, must be composed of materials with a very long stable lifetime. An important part of the R&D for a uniship is the development and use of age tolerant materials. The examination of materials used hundreds of thousands of years ago such as tools and dwellings shows the ravages of age. A uniship should have an initial goal of tens of millions of years of stability without aging. Ultimately one would hope that uniships that don't age in trillions of years could be built to travel to other universes.

These design requirements are far ahead of current technology.

9.3 Long-Lived Machinery and Electronics

If one has materials that preserve their composition, shape and performance characteristics over millions of years or more, then one can construct machinery and electronic gear such as computers that can last similar periods of time. Long-lived machinery and electronic gear then can be used when a uniship travels the multiverse.

---

[46] Neglecting the time required to accelerate to the starship's speed at the beginning and the time required to decelerate back to a "normal" speed of a few miles per second.

Long distance uniships need materials and machinery that last "nearly forever" – exactly the opposite of the intent of Earth industries.

9.4 Long Shelf Life Nuclear Reactors and Nuclear Shuttles

Several types of nuclear reactors are required for a uniship:

1. A continuous running reactor that can run for up to hundreds of millions of years to provide power to a uniship in flight to a distant location. This reactor may be a low power reactor. It should have a very long lifetime. That this is possible is suggested by the natural nuclear reactor that ran for millions of years about a billion years ago in the Congo.[47]
2. Long shelf life reactors that are not activated until a uniship destination is reached. These would power the uniship inside universes and their solar systems, and nuclear shuttles for travel and landings within a solar system.

9.5 Suspended Animation for Long Trips

It is necessary to have suspended animation available for crews on uniship journeys. With suspended animation a crew could go on a journey lasting gundreds of millions of years, or more, of uniship time, and, upon return to earth, have aged physiologically only a short time while out of suspended animation exploring distant universes and their star systems. The round trip travel time will not have aged them. When they return to earth they may be some months older, but their families and friends (having aged only by the earth travel time) will still be roughly contemporary with them.

A mechanism for long term suspended animation is thus a major requirement. Any suspended animation mechanism must take account of three important facts: 1) suspended animation must reduce human body temperatures to a low value to "halt" life processes and bodily decay; 2) lowering body temperatures will cause cells to rupture due to the expansion of water upon freezing; 3) the entry into suspended animation and the reentry to a normal bodily state must be rapid and uniform throughout the body.[48]

---

[47] The author suspects that the vast diversity of life in Africa may in part be due to genetic changes caused by this natural reactor. The development of mammalian life may in part also be due to this reactor and the radiation from its waste products and radioactive deposits in the Congo region over the millennia.

[48] One cannot "unfreeze" part of a human body and have the rest still frozen.

A mechanism to achieve these goals is not presently known. The current approaches to suspended animation (which all include lowering body temperature) are:

1. Replacing part or all of the blood in an organism with an "antifreeze" solution that will prevent cells and body tissue from bursting when the temperature is lowered. Revival takes place by raising the temperature of the organism while returning blood to the organism's circulatory system. This approach has been successfully applied to dogs that have been put into suspended animation for three hours. Unfortunately some of the dogs had nerve and coordination problems after revival.[49]

2. An organism can have a chemical injected or absorb a chemical while breathing that will counteract the tendency of water to expand when body temperature is lowered and/or lower the metabolic rate of the organism.

3. NASA and other groups have studied the possibility of placing humans into hibernation. Since hibernating organisms do age – perhaps more slowly – this approach is not true suspended animation.

4. A combination of electromagnetic "vibration" of a body having an innocuous chemical dispersed in the body (while awake or in suspended animation) might allow bodily temperatures to be lowered to a stable "frozen" state without cell rupturing. Turning off the electromagnetic vibration combined with a revival jolt might be an effective way to exit suspended animation procedure.

9.6 Robotic Driven Uniships

The initial uniship flights could be manned by robots rather than humans. This approach would be useful to test uniships, and their components, without endangering a crew. The robot guidance systems would, of course, have to be constructed of long-lived components. If it is successful then a robotic trip would also help demonstrate the long term reliability of long-lived computer equipment.

---

[49] At the University of Pittsburgh's Safar Center for Resuscitation Research.

Robotic flights would be especially useful if a method of rapid faster than light, or instantaneous, communication between the uniship and earth existed. (See chapter 10.)

9.7 Long-Life Computer Chips

Computers hardened for battle and bad weather conditions currently exist. A long distance uniship would require computers with working lifetimes of between thousands and hundreds of millions of years. In time periods of these lengths computer chips would be subject to aging processes such as the intermixing of the metals composing the various chips of the computer and the aging of the wiring of the computer. Since new materials of greater strength and other superior properties are being discovered fairly frequently one can hope that the required types of metals and materials will eventually be found.

9.8 Space Dust

The effect of dust and gas molecules in space on uniships are of great importance. These effect should be detectable in "short" distance uniship voyages. If it is important, as it seems to be, then the design of shielding for long distance uniships should incorporate appropriate "armor" to protect the uniship and crew.

9.9 Length Dilation Effects

Lengths on a uniship, traveling at high speed much greater than the speed of light, are significantly dilated. A length measured on a uniship will appear to be larger to an observer on earth by a factor of the speed measured in terms of the speed of light than the length on earth. For example if a uniship is moving at 5,000,000 times the speed of light then a two meter long stick (earth length) would appear to be 10,000,000 meters long on the uniship to an earth astronomer viewing the stick on the uniship.[50]

Does this length dilation phenomenon affect the contents of the speeding uniship? No. It is an illusion that the earth observer "sees." An occupant of the uniship would not notice a difference and would see the stick as still two meters in length.

---

[50] This discussion assumes that the stick and the starship motion are parallel. For a detailed discussion of this length contraction phenomena see Blaha (2011c) p. 17.

10 Multiverse Communications

10.1 Rapid Interstellar Communication

Recent work on quantum entanglement suggests that instantaneous communication may be possible using this mechanism if an advanced long range form of this laboratory phenomenon can be developed. Quantum entanglement can transcend the borders of universes since it is based on coordinated parts of a quantum state. Its instantaneous nature, which has been verified to great accuracy in recent experiments, makes it the ideal mechanism for communication over trillions of light years. We will describe the application of this concept in more detail in the following sections.

10.2 "Instantaneous" Interstellar Communication

Once a uniship capability is achieved it will clearly necessitate a very rapid, if not instantaneous, means of communication. All electromagnetic means of communication are limited by the speed of light and are thus insufficient for multi-million light year communication. If neutrinos are tachyons (faster than light) then they could provide a communications channel except that neutrino detection is very difficult, not reliable, and would require massive detectors that would be an unacceptable addition to the mass of a uniship. More importantly, because neutrinos are extremely light particles their speed is not much more than the speed of light at best.

The only possible method appears to be a quantum entanglement mechanism – currently a subject of intense scientific interest. Based on current thinking about this form of quantum communication it will have the following very desirable features:

1. It is a 1:1 form of communication with no possibility of being intercepted by others.
2. It requires a small amount of power no matter what the distance.
3. It is instantaneous and thus gives direct real time communications over any distance – even trillions of light years.

If history is any guide, the development of inter-universe communications will be similar to the development of telecommunications over the past 150 years, but on a much longer development time scale. Thus we anticipate that it will begin with a primitive Morse code equivalent, and progress eventually to fast digital transmission of images and data. We anticipate bilateral switchboards initially that eventually will lead to communications with uniships beyond our universe in the Flatverse or other universes. Obviously this capability would be needed for exploration – particularly by the initial robot-driven uniships, and for communications between colonies and earth scientific or commercial reasons in other universes.

The basic mechanism will consist of a bilateral quantum entanglement setup that begins as two electrons[51] of opposite spins in a quantum state with total spin zero. Each electron is nudged into a magnetic bottle that does not affect their joint spin state.[52] One bottle is retained on earth; the other bottle is placed in a uniship. As the uniship travels the state of the electron spin within its bottle can be periodically sampled but without changing its state.[53] This can also be done on the earth based electron in its bottle. If either electron's spin is flipped the spin of the other electron flips instantaneously no matter what the distance. Thus instantaneous communication of one computer bit takes place.

Eight such bottle pairs allow us by flipping bits to exchange bytes of data. Because of the time contraction associated with much faster than light uniships the byte change must be almost instantaneous for effective communication between a uniship and earth. This fast exchange can be done by ultrafast computers.

Eventually arrays of "bottles" can transmit bytes in bulk in support of large data and image transfer. One can envision electronic switchboards eventually linking arrays of bottles to form a network with a set of uniships and/or colonies. The thought processes and designs are similar to those used in telecommunications.

It is important to note that quantum communications does not require powerful transmitters Thus quantum communications is energy efficient.

---

[51] Protons would be another reasonable alternative.
[52] Several experimental groups have recently been able to detect parts within quantum states without affecting the overall quantum state.
[53] 2012 Nobel Prize winner Serge Haroche of France developed ways of detecting the state of particles without disturbing their quantum state.

10.3 Experimental Support for Instantaneous Quantum Entanglement Data Transfer

One might ask if instantaneous quantum data transfer is possible. Both quantum theory and numerous experiments have shown that instantaneous data transfer via entangled pairs works at large distances.[54]

10.4 Interstellar Communications and SETI

If our (and others) suggestion that quantum communication is the only reasonable way for communications at large distances, then this might be the reason for SETI's failure to find communications by alien civilizations. Aliens may very well not be communicating by radio or laser waves.

It is important to note that quantum communication, as we have proposed it, is inherently private 1:1 communication with no visible manifestations for others to detect of which we are aware.

---

[54] Matson, John, Quantum Teleportation Achieved Over Record Distances, Nature, 13 August 2012.

11   Uniship Development Time Frame

The development and deployment of uniships for the exploration of the Cosmos will be the supreme technological accomplishment of the human race. It will lead to an extraordinary growth of human culture to embody a combination of the best attributes of humanity enriched by contact with the cultures of the many extraterrestrial civilizations scattered across the multiverse.

11.1 The Development Phases

Although it is difficult to forecast how a complex development project will take shape, especially when so many parts of it require new technology in many areas, we shall make a tentative schedule of phases knowing that chance and difficulties will probably cause the actual schedule to differ. In making this schedule we will assume that major unforeseen breakthroughs will not occur. If a major breakthrough occurs such as "warp drive" then the development schedule would change drastically. However the author believes breakthroughs of that sort will not happen. Rather it is more likely to be a long term, expensive, hard slog towards eventual success.

We expect the following major phases to occur:

1. Development of efficient, large scale cargo and people transport to earth orbit over the next twenty years.
2. Manned inner planet exploration in the following thirty years using nuclear and possibly fusion powered space ships.
3. Manned exploration of the outer planets and moons and possibly the development of colonies on Mars and outer planet moons for scientific and commercial purposes. The time for this phase is probably another forty years following (but overlapping) the second phase.
4. Concurrent development of faster than light starships over a period of 150 years together with suspended animation and other technologies. The author believes the only viable propulsion approach is a quark-gluon ion drive described in Blaha (2013a) and earlier books. This approach would enable starships to evade the speed of light limit.

5. Assuming success in phase 4 an exploration and colonization phase of stars within a hundred light years of earth would follow with a duration of 1,000 years.

6. Technological advances in starships should then make exploration of the galaxy and nearby galaxies possible. This phase could last for many thousands of years and probably be extended to tens of thousands of years if economically and scientifically justified. Starships would continue to be propelled by quark-gluon drives using fusion energy.

7. Concurrent with phase 6 uniship design and development should take place using energy from particle-antiparticle annihilation – the most efficient energy source known. The many parts of this development will probably require many tens of thousands of years to put together. We suggest a ballpark figure of 50,000 years. A problem of major importance is to develop a method to reach extraordinary speeds of the order of millions of times the speed of light with uniship occupants experiencing humanly bearable accelerations.

8. After the design, development and testing of a uniship within our universe and in the nearby Flatverse an exploration program for the multiverse can commence with the exploration of "nearby" universes. This exploration phase will undoubtedly take tens of thousands of years up to millions of years depending on the benefits derived from exploration, and from perhaps meeting other civilizations.

11.2 Spaceship, Starship and Uniship Propulsion Phases

The different "ages" of space travel are best characterized by propulsion mechanisms and energy sources just as transportation on earth can be characterized by animal power (horses, etc.), steam powered engines, oil-powered engines, and jet engines.

The ages of space travel can be characterized as

1. The age of chemically powered spaceships
2. The age of Nuclear/Fusion Powered Spaceships (under development)
3. The age of Quark-Gluon powered Starships (at an initial experimental stage at CERN in particle physics high energy ion-ion collisions)
4. The age of particle-antiparticle powered Uniships (a gleam in the eye of multiverse enthusiasts)

11.3 Cost Issues

The starship development program mentioned above is a necessary precursor to the uniship development effort. The basic initial framework for the uniship project is the starship development. Uniships have similar design needs for the most part but on a much larger scale. Uniships will need nuclear and/or fusion engines for maneuvering in the solar systems of other universes. They may also need quark-gluon ion drive engines for low speed (but much greater than light speed) travel within a universe. Lastly they will need the most powerful energy source that we know of: particle-antiparticle powered engines for travel across the vast distances between universes.

In Blaha (2013a) we estimated the cost of design and construction of a faster than light starship at a trillion current dollars spread over one hundred years. In view of the continuing financial problems of the United States and Europe it is likely that the development time will be longer – perhaps one hundred and fifty years.

The development and construction of a uniship can build on the knowledge gained in the starship program. However its much greater requirements, and size, suggest that its cost will be in the five to ten trillion dollar range in current dollars.

Undoubtedly the research and development efforts in these programs will have tremendous technical spinoff benefits which will enrich world technology just as the space programs have done in the past. And the successful exploration of first the stars and galaxies of our universe, and then other universes should yield an enormously valuable return on our investment.

Appendix A describes our space and starship design and dynamics for the expansion of Man into the solar system and the galaxies of our universe. Appendix B describes the uniship design, and major considerations, for trips into the multiverse.

So we look to the future with confidence in the eventual expansion of humanity into the multiverse.

A From Earth to Quark-Gluon Starships for our Universe

The vehicles for voyages to space have primarily been via rockets. Recently a new generation of privately built rockets have been developed which promise to provide a more cost effective in transporting cargo into space. In addition the US Navy has developed and is planning to deploy rail guns for combat. This type of gun and the earlier big guns like Big Bertha (a World War I German gun) are capable of being engineered into space guns to cheaply loft cargo into space. Blaha (2013a) discusses the possibilities of non-rocket cargo transport to space – *particularly its relative inexpensiveness*. We can summarize some of the possibilities for space technology discussed in Blaha (2013a). We will then describe starship travel within our universe.

Some Topics in Blaha (2013a)

MULTI-STAGE SPACE GUNS FOR COST-EFFECTIVE CARGO SHIPMENT TO NEAR SPACE

Single-Stage Space Guns
Multi-Stage Space Guns
> Basic Multi-Stage Space Gun
> Enhanced Multi-Stage Space Gun

Other Measures to Minimize Space Gun Costs
Moving Massive Amounts of Cargo Into Space

NUCLEAR ROCKETS: THE PATH TO THE PLANETSERROR! BOOKMARK NOT DEFINED.

History of Nuclear Rocket R&D
> Bimodal Nuclear Thermal Rockets

NERVA Program Overview
Proposal For Nuclear Rocket Mars Missions vs. Ultra-large Chemical Rockets
New Russian Nuclear Rocket Program

A.1 Exceeding the Speed of Light – Starship Dynamics

In standard sublight relativistic dynamics the speed of a massive object cannot exceed the speed of light if a force applied to the object is real-valued. In this section we will consider the case of a *complex-valued* force applied to an object (complex thrust) that causes the object to attain a complex velocity whose real part can exceed the speed of light.

Complex valued forces (tachyonic forces) have not been experimentally found in Nature as yet. However the motion of a particle inside a Black Hole is tachyonic. And solid experimental evidence of tachyons exists (chapter 6 of Blaha (2010a), and also in (2013a)). In Blaha (2010a) we showed that neutrinos and d type quarks are tachyons. Up-type quarks also have complex 3-momenta. The real part (and absolute value) of their speed can be accelerated to faster than light or have speeds below the speed of light. If we harness quarks to create rocket thrust, then we have a mechanism for complex-valued thrust that can power starships faster than the speed of light.

One type of starship engine creates thrust by creating a quark-gluon plasma that acts as an ion drive to accelerate the starship. Fig. A.1 has a schematic diagram of a four linear accelerator system that creates a large quark-gluon plasma for ion propulsion. We expect the accelerators will accelerate heavy ions such as U-238.

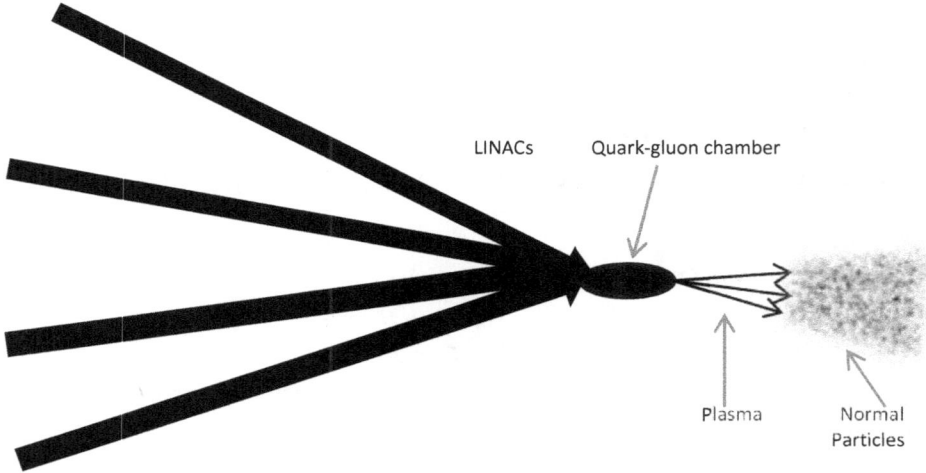

LINACs   Quark-gluon chamber

Plasma   Normal Particles

Figure A.1 Schematic of a linear accelerator engine with four linear accelerators at opening angles chosen to optimize the resultant quark-gluon thrust.

Fig. A.2 shows a design for a diamond-shaped starship using a four linear accelerator engine.

Since a complex-valued rocket thrust will generate a complex-valued velocity, and movement in space, the physical interpretation of complex velocities and distances must be addressed.

In appendix B of Blaha (2013a) we showed that a superluminal transformation maps points with real coordinate values in one coordinate system to points with complex coordinate values in the target coordinate system. We then showed that the complex-valued coordinate points in the target coordinate system could be "rotated" to real valued coordinates using $\Pi_L(\mathbf{v}/v)$ (eq. B.34). Thus a combined superluminal transformation and $\Pi_L(\mathbf{v}/v)$ transformation $\Pi_L(\mathbf{v}/v)\Lambda(\mathbf{v}/v)$ maps real coordinates to real coordinates. Complex coordinates are then merely an artifact of superluminal transformations.

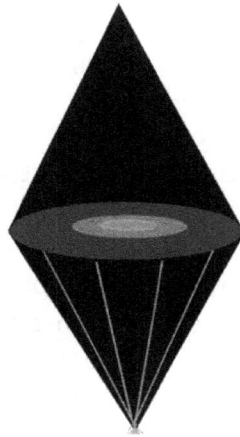

Figure A.2 A diamond shaped ship powered by four linear accelerators. The lower part has the accelerators, magnets, nuclear reactors, and propellant. The upper part contains the crew, cargo, and shielding as well as nuclear shuttles for exploring a solar system.

However when we consider the path of a rocket with complex thrust that starts from a spatial point with real coordinates and, as it accelerates, traverses complex-valued spatial points a new issue arises: What is the physical meaning of these complex-valued spatial coordinates. Unlike the previous case of superluminal transformations one cannot simply use a global transformation to change the complex-valued points to points with real coordinate values. This is particularly clear if one considers a three point configuration: the earth, the starship and the destination star.

The earth and star have real valued coordinates in the earth coordinate system. The starship in transit has complex coordinates at each point of its journey in starship coordinates. In general, there is no global transformation that will make the coordinates of all three points real-valued.

Therefore we conclude that complex coordinates are physically meaningful in this type of situation where one "point" is moving with a complex velocity. On this basis we will assume that space is three-dimensional with complex coordinate values in general. *But* using the *local* transformations of the Reality group described briefly in point 8 of appendix B of Blaha (2013a), namely local SU(3)⊗SU(2)⊗U(1)⊗SU(2)⊗U(1), these transformations can map complex

4-dimensional space to real-valued space – exactly the kind of space-time that we see. Thus we have a justification for Yang-Mills gauge theories – the type of gauge theories used in The Standard Model.

Why haven't the complex values of coordinates been noticed before? A simple, but the most important reason, for not noticing that coordinates are complex-valued, is that all space and time measurements use rulers and clocks that only measure real-valued numbers.

Additionally, we were unaware of complex-valued coordinates because objects with complex velocities have not been created and/or seen by Man. To give an object a complex velocity we need either a highly curved space-time region (such as a Black Hole) with an event horizon that encloses the object so that we can't see it; or we need to accelerate an object with a complex-valued force or thrust giving it a complex velocity and consequently a trajectory in complex coordinates.

This second possibility can only be achieved with tachyon thrust or force generated by accelerating quarks and gluons. Neutrinos, although tachyonic, only interact via the Weak interaction and have strictly real momenta. Thus they are not capable of generating a complex thrust.

Quarks are confined within protons and neutrons. To create macroscopic regions containing quarks we need collisions at enormous energies. We are just entering the experimental stage where this possibility can be realized. RHIC at Brookhaven National Laboratory and LHC at CERN have started creating quark-gluon fluids by colliding heavy ions such as gold and lead ions. Evidence for tachyonic quarks within the collision regions will hopefully be forthcoming soon.

Then a superluminal, tachyon drive starship with complex thrust is possible using quark-gluon ion drive.

A.2 Superluminal Starship Dynamics

In this section we will consider a constant, propulsive force in a starship's rest frame that drives the starship from a sublight velocity to a superluminal velocity. The key factor in achieving a superluminal speed is evading the singularity in γ at $v/c = 1$. We accomplish this goal by having a complex force – a force with a real and imaginary part – that generates a complex acceleration, and thus a complex velocity, that "goes around" the singularity in γ in the complex velocity plane. We assume that an "instantaneous" Lorentz transformation relates the earth reference frame and the starship reference frame at each point in time.

We assume the starship's thrust is in the direction of the positive x' (and x) axis. We also assume for simplicity that the mass of the starship is constant. (The starship engine uses a small amount of fuel relative to the starship's total mass.) The starship (primed coordinates) and earth (unprimed coordinates) coordinates have parallel axes. We assume the spatial force is in the positive x direction

$$\mathbf{F'} = g\hat{x} \tag{A.1}$$

where g is assumed to be a complex constant.

The fourth component of the force (since force is a Lorentz 4-vector) is zero in the rocket's rest frame:

$$F'^0 = 0 \tag{A.2}$$

Applying an inverse Lorentz transformation we find the force in the earth rest frame is

$$
\begin{aligned}
F^0 &= \gamma(F'^0 + \beta F'^x/c) = \gamma\beta F'^x/c = \gamma vg/c^2 \\
F^x &= \gamma(F'^x + \beta cF'^0) = \gamma F'^x = \gamma g \\
F^y &= F^z = 0
\end{aligned} \tag{A.3}
$$

where $\beta = v/c$, c is the speed of light, and $\gamma = (1 - \beta^2)^{-\frac{1}{2}}$ as before. We again use the superscripts x, y, and z to identify the components of the spatial force. The spatial momentum of an object of mass m is

$$\mathbf{p} = \gamma m\mathbf{v} \tag{A.4}$$

and the dynamical equation of motion in the earth's rest frame is

$$d\mathbf{p}/dt = \mathbf{F} \tag{A.5}$$

resulting in

$$dp^x/dt = \gamma g \tag{A.6}$$

with

$$dp^y/dt = dp^z/dt = 0 \qquad\qquad (A.7)$$

The differential equation resulting from eq. A.5 is

$$d(\gamma v)/dt = \gamma g/m \qquad\qquad (A.8)$$

assuming, as stated earlier, the fuel used is small[55] compared to the starship's mass. The solution of eq. A.8 is

$$v = c - 2c/(1 + ((c + v_0)/(c - v_0))\exp[2g(t - t_0)/(mc)]) \qquad (A.9)$$

where the velocity is v_0 at time t_0. If we take account of the decreasing fuel mass then we expect the velocity will increase somewhat more rapidly than eq. A.9. Since v is a complex number, due to the complex acceleration, the singularity at $v = c$ in γ is avoided and the starship can surpass the speed of light with no difficulty.

Integrating eq. A.9 we find the x distance traveled is

$$x = x_0 - c(t - t_0) + (mc^2/g)\ln((1 - v_0/c + (1 + v_0/c)\exp[2g(t - t_0)/(mc)])/2) \quad (A.10)$$

The complexity of g causes v and x to be complex. The complexity of x raises the question of its interpretation. The Reality group introduced in Blaha (2011d) and (2012a) furnishes the required interpretation. Since no spatial rotation takes place in this situation, and we are still using the earth coordinates, the relevant Reality group transformation only transforms the complex x value to a real value equal to its absolute value.[56] The matrix form of the transformation is

---

[55] Perhaps of the order of 10% – 20% of the starship mass.

[56] The Reality group appears in two roles. If a superluminal transformation is made between coordinate systems then a Reality group transformation is needed to make the target superluminal system coordinates into physical, real-valued coordinates such as an observer in that coordinate system would measure. Secondly if a dynamical state evolves in such a way that some of the coordinates become complex then a Reality group transformation is used to change complex coordinate values to physical, real-valued coordinates.

$$\Pi_L = \begin{bmatrix} 1 & 0 & 0 & 0 \\ 0 & e^{i\varnothing} & 0 & 0 \\ 0 & 0 & 1 & 0 \\ 0 & 0 & 0 & 1 \end{bmatrix} \qquad \text{(A.11)}$$

where $\varnothing$ is the phase of x. The real-valued coordinate x_p is the value of the distance traveled from earth which, of necessity, must be a real number measurable by a very long yardstick in principle.

$$x_p = |x| = xe^{-i\varnothing} \qquad \text{(A.10a)}$$

As a result superluminal travel to a distant star (or galaxy eventually) requires three phases in general. In the first phase (phase I) the starship accelerates with a value for the thrust g that enables it to reach a high complex velocity v_h. whose absolute value is much greater than the speed of light (eq. A.9). In the second phase (phase II) the starship coasts with speed v_h at a constant high speed to a point "not far" from the destination. In the third phase (phase III) the starship engines are turned on and the starship decelerates to a low speed at its destination solar system.

A.3 Achieving High Superluminal Starship Velocities

To achieve *much* faster than light motion the constant force value g required must satisfy a special set of conditions. These conditions emerge from a consideration of the denominator of eq. A.9:

$$1 + ((c + v_0)/(c - v_0))\exp[2g(t - t_0)/(mc)] \qquad \text{(A.12)}$$

If $v_0 < c$ then the denominator can only approach zero yielding an "infinitely large" velocity if g is a complex number. If $v_0 > c$ then the denominator can approach zero yielding an "infinitely large" velocity if g is a real or a complex number. The following cases are of interest:

<u>$c < v_0$</u>

In this case g is real and the zero of the denominator is specified by

$$2g(t - t_0)/(mc) = \ln((v_0 - c)/(c + v_0)) < 0$$

implying that g is negative and real since $t - t_0 > 0$.   A negative g value corresponds to deceleration of the starship if it is traveling faster than light.

c > v_0

In this case the denominator is zero if

$$2g(t - t_0)/(mc) = \ln((v_0 - c)/(c + v_0)) = \text{a complex number}$$

implying g must be complex. This case corresponds to a starship accelerating from a small speed to "infinitely large" speed.

We will consider the general case of complex g since this type of thrust allows us to go from very low speed to much faster than light speed – the general goal of starship travel. Let

$$g = g_1 + ig_2 \qquad\qquad (A.13)$$

If we wish the velocity to get very large (approach infinity) after some acceleration time interval $\triangle t = t_1 - t_0$ we set

$$1 + ((c + v_0)/(c - v_0))\exp[2g\triangle t/(mc)] = 0 \qquad\qquad (A.14)$$

with the result

$$g_2 = (mc/(2\triangle t))\{n\pi + \text{Im } \ln[(c - v_0)/(c + v_0)]\} > 0 \qquad (A.15)$$

and

$$g_1 = (mc/(2\triangle t)) \text{ Re } \ln[(c - v_0)/(c + v_0)] < 0 \qquad\qquad (A.16)$$

where n is an odd, positive integer, since v_0 is complex in general. Eqns. A.15 and A.16 enable the real and imaginary parts of the velocity (and thus the absolute value of the velocity) to become infinite as the time interval approaches $\triangle t$. We assume n = 1 in the following discussions. Substituting in eq. A.9 we obtain

$$v = c\{1 - 2/[1 + ((c + v_0)/(c - v_0))^{1 - (t - t_0)/\triangle t} e^{in\pi(t - t_0)/\triangle t}]\} \qquad (A.17)$$

We will now approximate eq. A.9's denominator as it approaches zero. Letting t = $t_1 + \tau$ where τ is small, and letting $\triangle t = t_1 - t_0$ then eq. A.9 becomes

$$v = c\{1 - 2/(1 + ((c + v_0)/(c - v_0))\exp[2g(\triangle t + \tau)/(mc)])\}$$
$$= c\{1 - 2/(1 - \exp[2g\tau/(mc)])\}$$
$$\simeq c\{1 - 2/(1 - (1 + 2g\tau/(mc))\}$$
$$\simeq c\{1 + (mc/g)(1/\tau)\}$$
$$\simeq (g^*mc^2/|g|^2)(1/\tau) \tag{A.18}$$

The magnitude of the acceleration is

$$|v| \simeq (mc^2/(|g|\tau) \tag{A.18a}$$

Eq. A.18 shows

- For small negative τ both the real and imaginary parts of $v \rightarrow +\infty$ as $\tau \rightarrow 0$ from below.
- For small positive τ both the real and imaginary parts of $v \rightarrow -\infty$ as $\tau \rightarrow 0$ from above.

The singular behavior as $\tau \rightarrow 0$ from above or below requires some explanation. It is not like the singular behavior as $v \rightarrow c$ seen in Special Relativity. Rather, it is a result of the use of the time coordinate of the earth's coordinate system. For if we transform earth time t to starship time t_r'' using

$$t_r'' = i\gamma(t - \beta x/c)$$

and determine the time contraction of an interval T from

$$T = \gamma_s T'' = (\beta^2 - 1)^{-\frac{1}{2}}T''$$

then we see that the starship time interval is

$$T'' \approx vT/c \tag{A.19}$$

Thus as v approaches infinity the starship time interval T'' grows to infinity as well. A starship will never reach the singular point $\tau = 0$ in a finite time. But, depending on fuel availability, it can reach speeds much, much faster than the speed of light.

The acceleration of the starship in the starship's coordinate system is

$$a' = F'^x/m = g/m \tag{A.20}$$

while the acceleration, a, in the earth's coordinate system is given by the derivative of eq. A.9.

$$\begin{aligned}
a &= dv/dt \\
&= 4(g/m)((c + v_0)/(c - v_0))\exp[2g(t - t_0)/(mc)]/\{1 + \\
&\quad + ((c + v_0)/(c - v_0))\exp[2g(t - t_0)/(mc)]\}^2
\end{aligned} \tag{A.21}$$

At $t = t_1 + \tau$ we see

$$a \simeq -mc^2/(g\tau^2)[1 + 2g\tau/mc] \approx -mc^2/(g\tau^2) \tag{A.22}$$

in the earth's reference frame. Thus the starship appears to have its acceleration approach infinity as τ approaches zero in the earth's coordinate system. This apparent problem does not exist in the starship reference frame where the acceleration is constant. Instead the starship takes an "infinite" starship time to reach infinite acceleration as seen on earth.

The crew can experience accelerations up to four times earth's normal gravitational acceleration without harm. If they are in suspended animation they should be able to withstand higher accelerations – perhaps eight times earth's gravitational acceleration.

The fuel expended as the earth time interval τ approaches zero in this case must approach infinity since there is a constant acceleration for an infinite time in the starship coordinate system. The acceleration is generated by the propellant exhaust. In this case we assume the fuel expended to accelerate to a high velocity is small compare to the starship's total mass.

We will examine the case where the fuel is not a small fraction of the starship mass later in this appendix.

A.4 Constant Superluminal Starship Travel

Assuming the starship has accelerated to an enormous *real* speed such as a speed between 5000c and 30,000c we can turn off the superluminal engines. The starship then moves at this constant speed in the absence of other forces, gravity, retarding effects of space dust, and so on.)

Consider a starship speed of 5000c. Any place in the galaxy is a short travel time away. And nearby galaxies are reachable as well. Fig. A.3 shows the time required to reach various interesting destinations at a much higher speed of 30,000c.

Destination	Distance (ly)	Approximate Travel Time (years)
To the other end of the Milky Way Galaxy	100,000	3
To the Center of the Milky Way	30,000	1
Large Magellenic Galaxy	150,000	5
Small Magellenic Galaxy	200,000	7
Andromeda Galaxy	2,000,000	70

Figure A.3. "Coasting" part of travel time to various destinations at a real velocity of 30,000c.

Since much, much higher "coasting" velocities are also possible almost the entire visible universe becomes accessible to Mankind if we can boost quark-gluon exhaust velocities to very large values. Mankind then has an incredible future if it has the will to seize it.

A.5 Deceleration of a Tachyonic Starship to Sublight Speeds
Eventually all journeys end, so we will now examine the deceleration of a starship as it approaches its destination. We turn on the superluminal engine. The thrust is reversed (g → –g) to decelerate as the target star system is approached.

A.6 Fuel Consumption
The acceleration of a rocket of mass m with a propellant exhaust speed v_e in the rocket's rest frame is given by

$$dv'/dt' = (v_e/m) \, dm/dt' \qquad\qquad (A.23)$$

and thus the constant g of eq. A.1 is

$$g = mdv'/dt' = v_e \, dm/dt' \qquad\qquad (A.24)$$

Since we intend to generate the thrust with a quark-gluon plasma producing an extremely high-energy exhaust we will *choose* the value of the starship acceleration to be equal to the acceleration due to gravity at the earth's surface g_E times $8(1 + i)$:[57]

$$g/m = 8(1 + i)g_E = 8(1 + i)980 \text{ cm/sec}^2 \qquad (A.25)$$

where m is the mass of the starship. If we specify an exhaust velocity v_e

$$v_e = -1000(c + ic) \qquad (A.26)$$

which is a reasonable choice for the exit speed thrust of the fireball then

$$dm/dt' = -2.61 \times 10^{-10} \text{ m} \qquad (A.27)$$

If the starship weighs 10,000 metric tons[58] then

$$dm/dt' = -2.61 \text{ gm/sec} \qquad (A.28)$$

From the viewpoint of rockets, dm/dt' is a small quantity. But due to time dilation the cumulative effect of dm/dt' in multi-year travel in starship time is a relatively large amount of fuel.

However if the starship can use processed material from asteroids and moons to make fuel then the limitation on travel imposed by fuel consumption can be circumvented. The fuel need not be composed of specific elements such as lead or uranium but could be spherules composed of a variety of materials if the starship engine were designed to handle such a variety of spherules.

The amount of fuel used per unit time would appear to be acceptable for quark-gluon plasma production for an ion drive. Currently minuscule amounts of plasma are created with ion-ion collisions. Colliding spherules of 1 milligram mass would require a not unreasonable collision rate of 1305 nominal collisions per second.

---

[57] Eight g's in astronaut terminology.
[58] About one-fifth the mass of the ship Queen Elizabeth.

We have seen that a starship with tremendous capabilities for exploring the universe can be built if we can build a quark-gluon ion drive that produces large complex accelerations.

B   Uniship Particle Annihilation Drive

> Speakin' in general, I 'ave tried 'em all—
> The 'appy roads that take you o'er the world.
> Speakin' in general, I 'ave found them good
> For such as cannot use one bed too long,
> But must get 'ence, the same as I 'ave done,
> An' go observin' matters till they die.
>   "Sestina of the Tramp-Royal" - Rudyard Kipling

Having developed a concept for starships that is arguably feasible within 150 years[59] if some major research and development efforts succeed, we now turn to the larger goal of designing a uniship. Fortunately, many of the features of a starship are features of a uniship on a much larger scale. In this appendix we will examine the major enhancements and additions needed for a uniship able to travel the multiverse.

B.1 Form of the Uniship Drive

The uniship drive that we foresee for that time some tens of thousands of years hence is based on the following premises, which may or may not be true:

1. The Baryonic force exists.
2. The multiverse exists, more or less, along the lines we have stated.
3. We can develop starships along the lines of Appendix A and our earlier work summarized in Blaha (2013a).
4. We can extend the features of starships to the magnitude of the uniship's much larger requirements.

The demands of traveling trillions of light years require the most energetic source of propulsion. Therefore it appears the uniship engine design will implement the following sequence of energy stages:

---

[59] Other approaches based on gravity waves and so on would not be feasible for tens of thousands of years.

1. Particle-antiparticle reactor that produces radiation – primarily gamma rays.
2. The radiation will be absorbed and transformed into heat and then electricity.
3. The electricity will power magnets and other electromagnetic parts of extremely high energy linear particle accelerators.
4. The confluence of the particle beams will produce a large quark-gluon plasma globule with quarks and gluons having extremely high complex-valued momenta.
5. Powerful magnets and the beam array geometry will cause a stream of faster than light ions to provide thrust for the uniship.

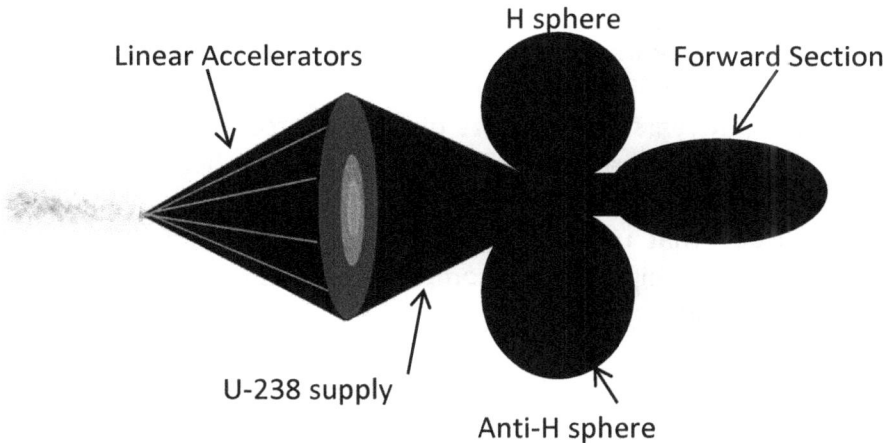

Figure B.1 A Uniship powered by a quark-gluon ion drive generated by the collision of four linear accelerator beams of heavy ions such as U-238. The spheres contain protons and antiprotons (or hydrogen and anti-hydrogen) respectively. The ship contains a supply of U-238 to fuel the linear accelerators. The forward section has the crew, cargo, electronics, and smaller nuclear powered craft to make trips within solar systems and land on planets.

Fig. B.1 schematically depicts a possible overall configuration of a uniship. The demands of extremely long distance travel with extremely long travel times in the uniship's reference frame place stringent requirements on uniship design. See chapter 9 for a detailed discussion.

B.2 Fuel, Acceleration, Speed, and Uniship Size Requirements

The fuel, acceleration, speed and uniship size requirements for travel in the multiverse are substantially larger than those required of a starship for travel in the universe. Distance scales in our universe are measured in millions of light years. Distance scales in the multiverse are probably of the order of trillions of light years. A heuristic argument for this larger scale is to compare the size of typical galaxies to the typical separation of galaxies. Galaxies typically are hundreds of thousands of light years in width and the distance between galaxies typically are millions of light years. The order of magnitude estimated ratio is $10^6/10^5 = 10$.

Our universe appears to be 13.8 billion light years in width. If we take the ratio for galaxies as a guide then the order of magnitude distance between universes is 10^{11} light years from the size of our universe which we can round off to a trillion light years as the order of magnitude spacing between universes on average.

We conclude that uniships should have speeds of the order of 1,000 times the speed of starships to have competitive flight times. Near the velocity singular point we found

$$|v| \simeq (mc^2/(|g|\tau) \qquad\qquad (A.18a)$$

The uniship time interval T'' is related to earth time T by

$$T'' \approx vT/c \qquad\qquad (A.19)$$

Thus an increase in the ship speed $|v|$ by a factor of 1,000 implies an increase in the uniship acceleration time T'' by a factor of 1,000 as well – in the case we considered of a constant force in the uniship reference frame.

If the maximum acceleration that humans can withstand is about $g = 8g_E$ then from the example in section A.6 we find eq. A.24

$$g = mdv'/dt' = v_e\, dm/dt' \qquad\qquad (A.24)$$

constrains v_e dm/dt' and there is no choice but to have much longer starship acceleration times (a factor of 1,000), and much larger fuel consumption if v_e, the quark-gluon thrust speed is limited. The uniship will spend about 1,000 times longer accelerating and decelerating, but the increase in earth time will be negligible.

Since fuel consumption aboard the uniship in this example is constant, the amount of fuel consumed will increase by a factor of 1,000 unless the thrust exit speed can be substantially increased. In this case less fuel will be consumed.

With much larger fuel consumption, and therefore mass, a uniship will be substantially larger than a starship – a factor of thousands at least.

REFERENCES

Blaha, S., 1998, *Cosmos and Consciousness* (Pingree-Hill Publishing, Auburn, NH, 1998).

_____2004, *Quantum Big Bang Cosmology: Complex Space-time General Relativity, Quantum Coordinates*™ *Dodecahedral Universe, Inflation, and New Spin 0, ½, 1 & 2 Tachyons & Imagyons* (Pingree-Hill Publishing, Auburn, NH, 2004).

_____ *2005a, Quantum Theory of the Third Kind: A New Type of Divergence-free Quantum Field Theory Supporting a Unified Standard Model of Elementary Particles and Quantum Gravity based on a New Method in the Calculus of Variations* (Pingree-Hill Publishing, Auburn, NH, 2005).

_____, 2005b, *The Metatheory of Physics Theories, and the Theory of Everything as a Quantum Computer Language* (Pingree-Hill Publishing, Auburn, NH, 2005).

_____, 2005c, *The Equivalence of Elementary Particle Theories and Computer Languages: Quantum Computers, Turing Machines, Standard Model, Superstring Theory, and a Proof that Gödel's Theorem Implies Nature Must Be Quantum* (Pingree-Hill Publishing, Auburn, NH, 2005).

_____, 2006, *A Derivation of ElectroWeak Theory based on an Extension of Special Relativity; Black Hole Tachyons; & Tachyons of Any Spin*. (Pingree-Hill Publishing, Auburn, NH, 2006).

_____, 2007b, *The Origin of the Standard Model: The Genesis of Four Quark and Lepton Species, Parity Violation, the ElectroWeak Sector, Color SU(3), Three Visible Generations of Fermions, and One Generation of Dark Matter with Dark Energy* (Pingree-Hill Publishing, Auburn, NH, 2007).

_____, *2008a, A Direct Derivation of the Form of the Standard Model From GL(16) (Pingree-Hill Publishing, Auburn, NH, 2008).*

_____, 2008b, *A Complete Derivation of the Form of the Standard Model With a New Method to Generate Particle Masses Second Edition* (Pingree-Hill Publishing, Auburn, NH, 2008)

_____, 2009a, *Bright Stars, Bright Universe* (Pingree-Hill Publishing, Auburn, NH, 2009)

_____, 2009b, *To Far Stars and Galaxies: Second Edition of Bright Stars, Bright Universe*   (Pingree-Hill Publishing, Auburn, NH, 2009).

_____, 2009c, *The Algebra of Thought & Reality: The Mathematical Basis for Plato's Theory of Ideas, and Reality Extended to Include A Priori Observers and Space-Time Second Edition* (Pingree-Hill Publishing, Auburn, NH, 2009).

_____, 2010a, *Operator Metaphysics: A New Metaphysics Based on a New Operator Logic and a New Quantum Operator Logic that Lead to a Mathematical Basis for Plato's Theory of Ideas and Reality* (Pingree-Hill Publishing, Auburn, NH, 2010).

_____, 2010b, *The Standard Model's Form Derived from Operator Logic, Superluminal Transformations and GL(16)* (Pingree-Hill Publishing, Auburn, NH, 2010).

_____, 2011a, *21st Century Natural Philosophy Of Ultimate Physical Reality* (McMann-Fisher Publishing, Auburn, NH, 2011).

_____, 2011b, *All the Universe! Faster Than Light Tachyon Quark Starships & Particle Accelerators with the LHC as a Prototype Starship Drive Scientific Edition* (Pingree-Hill Publishing, Auburn, NH, 2011).

_____, 2011c, *From Asynchronous Logic to The Standard Model to Superflight to the Stars* (Blaha Research, Auburn, NH, 2011).

_____, 2012a, *From Asynchronous Logic to The Standard Model to Superflight to the Stars volume 2: Superluminal CP and CPT, U(4) Complex General Relativity and The Standard Model, Complex Vierbein General Relativity, Kinetic Theory, Thermodynamics* (Blaha Research, Auburn, NH, 2012).

_____, 2012b, *Standard Model Symmetries, And Four And Sixteen Dimension Complex Relativity; The Origin Of Higgs Mass Terms* (Blaha Reasearch, Auburn, NH, 2012).

_____, 2013a, *Multi-Stage Space Guns, Micro-Pulse Nuclear Rockets, and Faster-Than-Light Quark-Gluon Ion Drive Starships* (Blaha Reasearch, Auburn, NH, 2013).

_____, 2013b, *The Bridge to Dark Matter; A New Sister Universe; Dark Energy; Inflatons; Quantum Big Bang; Superluminal Physics; An Extended Standard Model Based on Geometry* (Blaha Reasearch, Auburn, NH, 2013).

_____, 2014, *Universes and Multiverses: From a New Standard Model to a Physical Multiverse; The Big Bang, and Our Sister Universe Wormhole; Cosmological Constant's Origin; A Baryonic Field between Universes and*

Particles; Flatverse Extended Wheeler-DeWitt Equation (Blaha Reasearch, Auburn, NH, 2014).

Eddington, A. S., 1952, *The Mathematical Theory of Relativity* (Cambridge University Press, Cambridge, U.K., 1952).

Fant, Karl M., 2005, *Logically Determined Design: Clockless System Design With NULL Convention Logic* (John Wiley and Sons, Hoboken, NJ, 2005).

Gradshteyn, I. S. and Ryzhik, I. M., 1965, *Table of Integrals, Series, and Products* (Academic Press, New York, 1965).

Rescher, N., 1967, *The Philosophy of Leibniz* (Prentice-Hall, Englewood Cliffs, NJ, 1967).

Sakurai, J. J., 1964, *Invariance Principles and Elementary Particles* (Princeton University Press, Princeton, NJ, 1964).

Streater, R. F. and Wightman, A. S., 2000, *PCT, Spin, Statistics, and All That* (Princeton University Press, Princeton, NJ 2000).

Weinberg, S., 1995, *The Quantum Theory of Fields Volume I* (Cambridge University Press, New York, 1995).

Weyl, H., 1950, *Space, Time, Matter* (Dover, New York, 1950).

Weyl, H., (Tr. S. Pollard et al), 1987, *The Continuum* (Dover Publications, New York, 1987).

INDEX

About the Author

Stephen Blaha is an internationally known physicist with interests in Science, the Arts, and Technology. He had an Alfred P. Sloan Foundation scholarship in college. He received his Ph.D. in Physics from Rockefeller University. He has served on the faculties of several major universities. He was also a Member of the Technical Staff at Bell Laboratories, a manager at the Boston Globe Newspaper, a Director at Wang Laboratories, and President of Blaha Software Inc and of Janus Associates Inc. (NH).

Among other achievements he was a co-discoverer of the "r potential" for heavy quark binding developing the first (and still the only demonstrable) non-abelian gauge theory with an "r" potential; first suggested the existence of topological structures in superfluid He-3; first proposed Yang-Mills theories would appear in condensed matter phenomena with non-scalar order parameters; first developed a grammar-based formalism for quantum computers and applied it to elementary particle theories; first developed a new form of quantum field theory without divergences (thus solving a major 60 year old problem that enabled a unified theory of the Standard Model and Quantum Gravity without divergences to be developed); first developed a formulation of complex General Relativity based on analytic continuation from real space-time; first developed a generalized non-homogeneous Robertson-Walker metric that enabled a quantum theory of the Big Bang to be developed without singularities at t = 0; first generalized Cauchy's theorem and Gauss' theorem to complex, curved multi-dimensional spaces; received Honorable Mention in the Gravity Research Foundation Essay Competition in 1978; first developed a physically acceptable theory of faster-than-light particles; first showed a universe with three complex spatial dimensions is icosahedral; first derived a composition of extrema method in the Calculus of Variations; first quantitatively suggested that inflationary periods in the history of the universe were not needed; first proved Gödel's Theorem implies Nature must be quantum; provided a new alternative to the Higgs Mechanism, and Higgs particles, to generate masses; first showed how to resolve logical paradoxes including Gödel's Undecidability Theorem by developing Operator Logic and Quantum Operator Logic; first developed a quantitative harmonic oscillator-like model of the life cycle, and interactions, of civilizations; first showed how equations describing superorganisms also apply to civilizations; and first developed an axiomatic derivation of the forms of The Standard Model with DARK PARTICLEs from geometry – space-time properties – The faster than light Standard Model.

He has had a major impact on a succession of elementary particle theories: his Ph.D. thesis (1970), and papers, showed that quantum field theory calculations to all orders in ladder approximations could not give scaling deep inelastic electron-nucleon scattering. He later showed the eigenvalue equation for the fine structure constant α in Johnson-Baker-Willey QED had a zero at $\alpha = 1$ not 1/137 by solving the Schwinger-Dyson equations to all orders in an approximation that agreed with exact results to 8th order in α thus ending interest in this theory. In 1979 at Prof. Ken Johnson's (MIT) suggestion he calculated the proton-neutron mass difference in the MIT bag model and found the result had the wrong sign reducing interest in the bag model. These results all appear in Physical Review papers. In the 2000's he repeatedly pointed out the shortcomings of SuperString theory and showed that The Standard Model's form could be derived from space-time geometry by an extension of Lorentz transformations to faster than light transformations. This deeper space-time basis greatly increases the possibility that it is part of THE fundamental theory.

In the early 1980's Blaha was also a pioneer in the development of UNIX for financial, scientific and Internet applications: benchmarked UNIX versions showing that block size was critical for UNIX performance, developing financial modeling software, starting database benchmarking comparison studies, developing Internet-like UNIX networking (1982) and developing a hybrid shell programming technique (1982) that was a precursor to the PERL programming language. He was also the manager of the AT&T ten-year future products development database. His work helped lead to commercial UNIX on computers such as Sun Micros, IBM AIX minis, and Apple computers.

In the 1980's he pioneered the development of PC Desktop Publishing on laser printers. and was nominated for three "Awards for Technical Excellence" in 1987 by PC Magazine for PC software products that he designed and developed.

In the past ten years Dr. Blaha has written over 35 books on a wide range of topics. Some recent major works are: *From Asynchronous Logic to The Standard Model to Superflight to the Stars, All the Universe!* and *SuperCivilizations: Civilizations as Superorganisms.*